D1205613

NATURAL WINE

2010 VINE
SONOMA COUNTY

NATURAL WINE

an introduction to organic and biodynamic wines made naturally

ISABELLE LEGERON MW

CICO BOOKS
LONDON NEW YORK

TO BO'AH, FOR MAKING IT HAPPEN

This third edition published in 2020 by CICO Books
An imprint of Ryland Peters & Small Ltd
20–21 Jockey's Fields 341 E 116th St
London WC1R 4BW New York, NY 10029

www.rylandpeters.com

10 9 8 7 6 5 4 3 2

First published in 2014

For additional picture credits, see page 224.

A CIP catalog record for this book is available from the Library of Congress and the British Library.

ISBN: 978 1 78249 899 5

Printed in China

Editor: Caroline West
Designer: Geoff Borin
Photographer: Gavin Kingcome
Illustrator: Anthony Zinonos

Art director: Sally Powell
Head of production: Patricia Harrington
Publishing manager: Penny Craig
Publisher: Cindy Richards

Note to reader
Where possible, metric and imperial equivalents have been provided (for example, for measurements of distance), except in quoted material. Throughout the book, sulfite levels are given in grams per liter (with 1 liter being the equivalent of approximately 34 US fl. oz).

CONTENTS

We live in a society where it's fashionable to wear farmer's boots and the chit-chat de rigueur at the local butcher's revolves around how long your meat has been hung. Microbreweries and espresso bars populate our urban landscapes, and yet, even against this new agro-chic backdrop, we still, without the slightest thought, wash down our outdoor-reared sausages with the vinous equivalent of a battery chicken. Perhaps this is because, while it's become routine to look at the list of ingredients on the back of most foodstuffs, with wine, we can't, as no such labeling laws exist.

This book isn't meant to be an exposé of the wine world. Rather, it is a tribute to those wines that are not only farmed well, but also fly in the face of modern winemaking practices, remaining natural against all the odds. It is also a celebration of the remarkable people who create them. Like sailors going to sea, playing the winds and riding the waves, these winemakers understand that nature is much greater than themselves. They acknowledge that not only is it futile, but actually counterproductive, to try to control or tame her, as her magic lies in her power.

I am not a winemaker, nor do I pretend to know everything about the science of winemaking. However, I have an overall vision based on discussions with growers, as well as tasting and drinking thousands of wines. I always intended this book as a starting-point, an invitation to people to explore and begin asking their own questions. My personal views are clear and I don't sit on the fence. Apart from the fact that I genuinely believe all wine should be farmed organically as a bare minimum, there is no political or economic agenda behind the writing. Instead, my opinions are guided by what I enjoy drinking. I believe wines made naturally, with no (or very few) sulfites, taste the best, and this is why I drink nothing else. It is with this in mind that I wrote the book.

Natural Wine is therefore a subjective look at what makes great wine, because, for me, only natural wine can be truly great. I have tried to tell as much of the story as possible through the voices and stories of others, because this world is not my creation. It is real, it exists, and many of the thoughts and experiences I share are those of a much larger community. While doing my research, I found that there's very little written information on the subject, not least because most of the conventional wine world disregards the natural as not commercially viable. Consequently, my findings are largely based on primary research: conversations, interviews, and, of course, a lot of wine tasting.

Wine is something we ingest. Like other types of food, it can be more or less wholesome, more or less manipulated, and more or less delicious. In many ways, this book could easily apply to other foodstuffs, including bread, beer, and milk, that have suffered a similar over-commercialized fate (and natural revival); it's just that wine has been a little slow off the mark. So, if you understand how proper food can provide a nourishment that goes beyond merely satisfying hunger and that the energy, commitment, and intentions of natural wine producers matter, then you'll see just how special fine, natural wine is—and I hope you will never look back.

ISABELLE LEGERON MW

INTRODUCTION

FARMING TODAY

I recently spent a weekend with friends in a beautiful country house in Cornwall. As I watched the fields roll, wave-like, in the sea winds, it dawned on me that this idyllic setting was anything but. For miles, all I could see were cornfields growing on rock-hard, barren earth; not a single other plant was growing amid the green stalks. It was both shocking and extraordinary to see how, in an instant, the same gentle landscape could suddenly seem different, stark, and lifeless.

Nowadays, agricultural monoculture is so prevalent that we don't even notice it. From our neatly trimmed, dandelion-banished, perfect-green lawns to the vast expanses of cereals, sugar beet, and even grapes that blanket our countryside, we like to have nature under control. Where before you might have seen small pockets of pastureland, woodland, and crop fields, carved up by hedgerows that acted as wildlife motorways, today views are dominated by monotony. Since 1950 the number of farms in the United States, for example, has halved, while the average size of those remaining has doubled, so that today only two percent of the country's farms produce 70 percent of its vegetables.

Below: **Monoculture galore in California: miles of grapes and more grapes.**

The 20th century changed the face of agriculture. It streamlined, mechanized, and "simplified" farming in an attempt to increase yields and maximize short-term profits. This industrialization became known as the "Green Revolution." "We call it 'intensification,' but it was intensification per *farmer*, not per square meter," explain agronomists Claude and Lydia Bourguignon. "In North America, yes, a single farmer can manage 500 hectares alone, but the traditional agro-silvo-pastoral farming system was actually far more productive per square meter."

Grape-growing, like the rest of agriculture, is no exception. "Traditionally, in Italy, vines were very biodiverse," explained the late Stefano Bellotti, a natural grower in Piedmont. "They grew alongside trees or vegetables, and growers also cultivated wheat, beans, chickpeas, and even fruit trees, between the rows. Biodiversity was very important."

Modern agriculture has been about developing duplicable approaches that can be applied uniformly wherever you are. It is what natural Californian grower Mary Morwood Hart calls "textbook farming." Mary explains: "These consultants come around telling you how many leaves there should be per bunch of grapes without considering any of the particularities of your site." In fact, as Tony Coturri, a natural grower in Sonoma, puts it, the industry has become so mechanized and detached from its roots that not only has "most wine today never even seen a human hand," but also "the growers don't call themselves 'farmers.' They don't see viticulture or grape-growing as an agricultural pursuit." This approach couldn't be more different from that of growers like Sébastien Riffault, in Sancerre, France, who considers each vine individually. He says, "They're like people: each plant needs different things at different times."

Above: **Traditionally, grapes were harvested by hand. This unmechanized approach continues today in many vineyards driven by quality.**

One of the biggest causes of this disconnection seems to have been the development of synthetic chemical treatments (such as fungicides, pesticides, herbicides, insecticides, and fertilizers), which were all created to facilitate farmers' work, but which, inevitably, led to them stepping back from the needs of the *living* world in their care. The problem is that spraying herbicides or feeding plants with nitrogen-rich fertilizers, for example, doesn't start and end in the vineyard. It causes fundamental imbalances in the ecosystem, with some products leaking through into the groundwater. "This is the very start of the chain," says natural grower Emmanuel Houillon from the Jura, in eastern France. "There are even synthetic products that remain attached to water molecules during evaporation so that they fall with raindrops."

According to the World Wildlife Fund, the amount of pesticides sprayed on fields has increased 26-fold over the past 50 years. Vineyards, in particular, seem to have played a huge part, with the application of synthetic pesticides to European vineyards having increased by 27 percent since 1994, according to the Pesticide Action Network (PAN). PAN states that "Grapes now receive a higher dose of synthetic pesticides than any other major crop, except citrus."

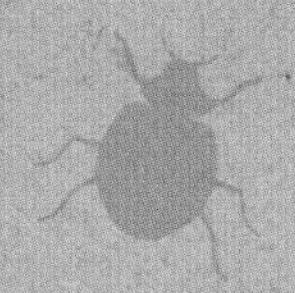

This has a detrimental effect on soil life, Claude and Lydia Bourguignon explain. "Soils harbor 80 percent of the world's biomass. Earthworms alone, for example, amount to about the same weight as all other animals combined. But, since 1950, European numbers have decreased from two tonnes per hectare to less than 100kg."

This biological degradation has a profound effect on the soil, eventually also leading to chemical degradation and massive soil erosion. "When agriculture began some 6,000 years ago, 12 percent of the planet was covered in desert; today, 32 percent is," continue Claude and Lydia. "And out of the two billion hectares of desert that we have created in this time, half of this was created in the 20th century." It is a yearly decline that is drastically diminishing our natural capital. "Recent estimates suggest that each year more than 10 million hectares of cropland will be degraded or lost, as wind and rain erode topsoil," explains ecologist and author Tony Juniper.

We are not separate from our environment and even less so from what we eat and drink. In fact, two separate studies by PAN (in 2008) and the French consumer organization UFC-Que Choisir (in 2013) found pesticide residues in the wines they tested. While the totals were tiny (measured in micrograms per liter), they were nonetheless significantly greater (sometimes more than 200 times higher) than the accepted standard for United Kingdom drinking water. Some of the residues were even carcinogenic, developmental or reproductive toxins, or endocrine disruptors. Given that wine is 85 percent water, it certainly makes you wonder.

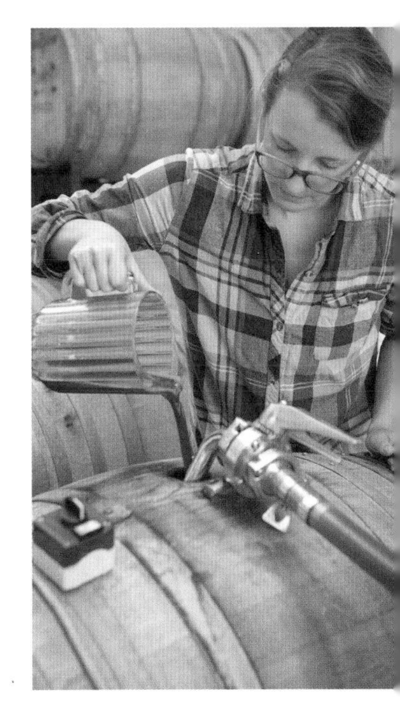

Above: **Natural wines require precision as well as a lot of care and attention from those producing them.**

Opposite: **A wild Californian vineyard where grapes rub shoulders with apple trees, brush, and native grasses.**

SOILS HARBOR 80 PERCENT OF THE WORLD'S BIOMASS. EARTHWORMS ALONE, FOR EXAMPLE, AMOUNT TO ABOUT THE SAME WEIGHT AS ALL OTHER ANIMALS COMBINED.

WINE TODAY

> "Wine is simple. Life is simple.
> It's man that complicates, and it's a shame."
> (BERNARD NOBLET, FORMER CELLAR MASTER, DOMAINE DE LA ROMANÉE CONTI, FRANCE)

Above: **Old, gnarled, indigenous vines like this are often grubbed because of low yields or for being unfashionable. They are, however, often the most adapted and intimately connected to the land, having developed deep root systems.**

In 2008, when I first traveled to Georgia, in the Caucasus, I was amazed to discover that almost every family makes wine at home and, if they have a surplus, they sell it for extra cash. Sure, some of what I tasted was lovely, some undrinkable, but what is of note is that for rural Georgians wine is simply a part of their diet. Just as they rear pigs to eat pork, grow wheat to make bread, and raise a cow or two for milk, they also grow grapes for wine.

While Georgia's subsistence farmers may be the exception to the rule nowadays, it was not always so. Wine started life everywhere as a simple drink, but then morphed over time to become a branded, consistent, standardized commodity, the production of which is primarily informed by the bottomline, while also being subject to the vagaries of fashion and consumerism. And what a shame that is.

It means that, often, farming decisions are made, not with the longevity of the plant or its environment in mind, but in terms of how quickly the producer can make a return on his or her investment. Vines are planted in places that they probably shouldn't be, farmed poorly, and then, once the grapes hit the cellar, dozens of additives, processing agents, and manipulations are used to *manufacture* a standardized product. Like so many other industries, wine moved from being handmade and artisanal to being large-scale and industrialized.

There is nothing particularly remarkable about this except that, unlike in other industries, our impression of how wine is made seems to have stayed put. People still believe that wine is produced by humble farmers with as little intervention as possible—and brands everywhere are happy to comply with this illusion. When you realize, for example, that three wine companies accounted for nearly half of all the wine sold in the United States in 2012,

while, in Australia, the top five accounted for over half of the national crush, it's clear that there is a disconnection between what wine is and what it *appears to be*.

Fair enough, you might think; after all, mergers and acquisitions are common practice nowadays. Plus, wine seems a pretty tricky thing to make: you need high-tech equipment, expensive buildings, highly trained individuals... And yet, you don't. Left to their own devices, organic compounds that contain sugar ferment naturally and grapes are no exception. Grapes are surrounded by living organisms that are ready to break them down, and one of the possible outcomes of this *natural* process is wine. Simply put, if you pick grapes and squash them in a bucket, you will, with a little luck, end up with wine.

Over the course of time, people perfected this bucket technique. They found places where, year in and year out, vines gave great grapes, and they developed methods to help them understand the magic that makes the grape–wine transformation possible. However, while advances in technology and winemaking science have been enormously positive for the industry as a whole, today we seem to have lost perspective.

Above: **Unlike most vineyards today, polyculture still plays an important role in natural wine production, as is the case at the Klinec farm in Slovenia.**

Above: **A vineyard producing natural wine in the Veneto, northern Italy.**

Opposite (below): **Not only are the same international grape varieties planted all over the world, but wine styles are also duplicated, resulting in, as Hugh Johnson remarks, "a homogenization of wine production, but, whereas it always used to be a case of the New World following Europe, now Europe is following the New World."**

Rather than use science to produce wines with as little intervention as possible, we use it to gain absolute control over every step of the process—from growing the grapes to making the wine itself. Very little is left to nature. Instead, most wine today, including expensive, so-called "exclusive" examples, is a product of the agrochemical food industry. And what is extraordinary is that most of this change has happened over the last 50-odd years.

It was also not until the second half of the 20th century that commercial selected yeast strains became available. Lallemand, for example, one of the world's leading suppliers (and manufacturers) of selected yeast and bacteria, only started selling wine strains in 1974 (in North America) and in 1977 (in Europe).

It is the same with other additives, such as the infamous sulfites, whose effect on wine is, as natural Champagne producer Anselme Selosse puts it, "much the same as Jack Nicholson in *One Flew over the Cuckoo's Nest*. It lobotomizes the wine." Contrary to popular wine-industry belief, the use of sulfites in winemaking (in order to keep barrels clean) is actually relatively recent, while their use as an additive mixed into wine is even more so (see *A Brief History of Sulfites*, pages 68–69).

Interventionist technologies are surprisingly recent, too, even though many are often used in wine today. "Sterile filtration is very modern," says Gilles Vergé, a natural grower in Burgundy, France. "It only started being used in my area in the 1950s and reverse osmosis [whose filter membrane is so tight that it is almost 10,000 times tighter than a sterile filter] at the end of the '90s." And, while the use of reverse osmosis (RO) is still pretty hush-hush, according to Clark Smith, a wine consultant credited with popularizing the process, far more machines are sold than producers admit.

As the recent family history of Montebruno's Joseph Pedicini, an ex-brewer-cum-natural-wine producer in Oregon, illustrates, these innovations are really very new. "In 1995, when I was still brewing, I tried taking over the home winemaking for my family (we're originally Italian and my grandparents brought their winemaking know-how over with them). I applied my brewing knowledge and introduced things like laboratory yeast strains. My relatives would look at me, scratching their heads:

'Why's he putting all that in our wine?!'

'Just a minute, Zio, I learned this in school, it's going to be good!'

But the wines came out soul-less. Tasty, but lacking the magic."

Whether it is Joseph's family in New Jersey or my rural Georgian folk, it comes back to the same thing. *Wine makes itself*.

Above: **Many wineries today have minimized the human element of winemaking by mechanizing production.**

"Natural wine is not new; it is what wine always was, and yet, somehow today it has become a rarity. It is a tiny drop in a big ocean, but, oh my, what a drop."

(ISABELLE LEGERON MW)

PART 1

WHAT IS NATURAL WINE?

Above: **A barrel sample of fermenting natural wine.**

Opposite: **Natural wine in the making from the 2013 harvest in the Beaujolais.**

Previous page: **Healthy vines growing at Mythopia, an experimental vineyard in Switzerland that produces a raft of natural wines.**

"IS THERE SUCH A THING AS NATURAL WINE?"

During the summer of 2012, inspectors from the Italian Ministry of Agriculture descended on Enoteca Bulzoni, a wine shop in Rome's Viale Parioli, which had been operating successfully since its inception in 1929. Owners Alessandro and Ricardo Bulzoni, grandsons of the founders, suddenly found themselves slapped with a fine, as well as a possible prosecution for fraud, for selling *vino naturale* (natural wine) without certification.

When questioned about their actions, Italian Ministry officials explained that the phrase "natural wine" did not legally exist. Whereas other denominations and labeling terms are subject to rules and regulations that impose constraints on their usage, natural wine is not because no certification currently exists. This, argued the Ministry, meant that it was not verifiable and could, therefore, be misleading to the public, as well as damaging to other producers who did not label their wines in this way. The brothers Bulzoni paid up and went straight back to selling the stuff.

The Italian daily newspaper *Il Fatto Quotidiano*, which covered the story at the time, summed up the conundrum. On the one hand, you had the Bulzoni brothers, whose family had been selling wine for three generations and had always had their clients' best interests at heart. They were making no claims to the "better-ness," or otherwise, of natural wine, but were, instead, simply using a common phrase to isolate wines that had been produced without the use of additives. On the other hand, you had the Ministry. While it agreed in principle that the "natural wines" may indeed have been produced without additives, it also insisted that the law had to be respected. And the law, as it currently stands, has no definition for natural wine.

This is one of the biggest problems faced by natural wine producers today. There is, as yet, no official accreditation for their product, which leaves the term open to abuse and thus to criticism. As Jem Gardener of Vinceremos,

> "Modern winemaking is a lot about using SO2, controlling fermentation and temperatures. But there is an alternative way, and it is better."
> (DAVID BIRD MW, CHARTERED CHEMIST AND AUTHOR OF *UNDERSTANDING WINE TECHNOLOGY*)

Above left: **Macerating berries undergoing their alcoholic fermentation. This process occurs spontaneously with healthy berries that have been farmed cleanly.**

Above right: **Red grape pomace left over after pressing. In organic vineyards, this is then typically used for mulching or composting.**

a specialist organic wine shop in Leeds, in the United Kingdom, explains, "We are expected to take it on trust... that they are using natural methods and ingredients. I'd love this to be sufficient, but I fear in this world it isn't." As things stand at the moment, any grower can call himself natural. Whether or not he is comes down to his own integrity.

INTERFERENCE—HOW MUCH IS TOO MUCH?

To complicate matters further, vinification is a tricky business and deciding how much manipulation is acceptable is far from simple. Some 50 additives and processing aids, for instance, can be used in the making of a European Union organic wine alone. While all natural wine producers would agree, for example, that adding flavor-enhancing yeasts is a no-go area, some would argue that the addition of a small quantity of sulfites at bottling stage is acceptable. Similarly, some believe that fining or filtering are manipulations that fundamentally change the structure of a wine and so should not be allowed. Others argue that traditional practices such as the use of organic egg whites for fining do not make their wines any less natural.

Given the confusion, an official definition seems both necessary and inevitable. As natural wines swell in numbers, along with the public's receptivity to them and questioning of what other types of wine have generally become, so too have the incentives to capitalize on the popularity of the "natural." Some larger producers have begun releasing so-called "natural" cuvées or using the term in marketing material to refer to wines that are definitely conventional. Whether this is done in genuine confusion or in a cynical attempt to capitalize on the trend and its positive connotations, the result is the same—confused drinkers.

DEFINING AND REGULATING NATURAL WINES

Authorities are beginning to respond. In the fall (autumn) of 2012, for instance, the *Association des Vins Naturels* (AVN), a French association of natural wine growers (see *Where and When: Grower Associations*, pages 120–21), met up with the fraud squad and other officials in Paris, to discuss the possibility of submitting a definition of their production methods for official registration so that it might be used to verify wines hitting the market as "natural." As Bernard Bellahsen, owner of Domaine Fontedicto and one of the AVN founders, explained, "They finally understood that there was a difference between organic and natural, so they asked us to fine-tune our definition and officially lodge a '*cahier des charges*'—or rulebook. It's pretty simple where they are concerned. They simply apply the law. As an association, you provide them with points of reference and they work out if the products correspond or don't. It just has to be registered and declared officially." While it remains vague, however, they can't run checks, which is why, as in Italy, the authorities are unhappy about the use of the term. In fact, in the fall of 2013, just before the first edition of this book went to print, the Italian government launched a parliamentary inquiry into "natural wine" in an attempt to provide greater clarity. To date it is still on-going.

What is certain is that, as a group, natural wines are growing. Perhaps it is this success that has engendered such controversy from within the ranks of some of the wine trade: "There's no such thing as natural" or "How dare they imply my wines aren't natural?"

In fact, some natural growers don't like the term "natural" either. "It's not a great word because it can be twisted in all sorts of directions," says Anne-Marie Lavaysse of Le Petit Domaine de Gimios, a vineyard in the Languedoc, in France. Bernard Bellahsen agrees: "When I talk about natural wine, I say 'it's just fermented grape juice. I use grapes, grapes, and more grapes, and the result is wine. That's it.' Admittedly, it isn't especially succinct, but it's truer because lots of things can be called "natural," making them seem healthier, when often it isn't the case. It's very delicate terminology."

So, yes, perhaps "natural wine" is not the best term to use. In fact, it's a shame that any adjective should have to be added to set apart a wine that is what dictionaries everywhere describe it to be. But, unfortunately, the world

Above: **As things currently stand there is no compulsory, legal inclusion on a label that enables drinkers to tell natural wine apart.**

has moved on and today wine doesn't just mean "fermented grape juice," but "grape juice fermented with X, Y, and Z," so the term "wine" has to be qualified to set these particular examples apart.

And, yes, perhaps a less contentious term, such as "live," "pure," "real," "true," "low-intervention," "authentic," "farmhouse," or some such, would be less provocative. But it is the term most commonly used globally to describe wines of this nature. For whatever reason, people all over the world have chosen to use the term "natural," in the face of all the alternatives, to describe healthily grown, nature-friendly, low-intervention wines that truly express their place of origin. As the late Piedmontese natural wine grower Stefano Bellotti, of Cascina degli Ulivi, put it, "I'm not thrilled about the term 'natural,' but that's the way it is. If you don't like the word 'table,' you can't just call it a 'chair.'" So, natural wine it is.

Whether or not it is certified (or indeed certifiable), natural wine does exist. It is wine from vineyards that are farmed organically, at the very least, and which is produced without adding or removing anything during vinification, apart from a dash of sulfites at most at bottling. This makes it the closest thing there is to the good, old, fermented grape juice that Google believes wine to be.

"Pick grapes and ferment" may sound easy and self-evident, but, once you start picking apart the reality, you soon realize that in its purest form natural wine is almost a miraculous feat—a great balancing act between life in the vineyard, life in the cellar, and life in the bottle.

Opposite and below: **Natural wines come from vineyards that nurture and protect life, from vine to cellar to bottle.**

THE VINEYARD: LIVING SOILS

> "The nation that destroys its soil destroys itself."
> (FRANKLIN D. ROOSEVELT, LATE PRESIDENT OF THE UNITED STATES)

When Dr. Grace Augustine, a botanist in James Cameron's film *Avatar* (2009), marches into the control room at Hell's Gate and lets rip at the Head of Mining Operations for bulldozing a hallowed, willow-like tree, I am sure that most people in the audience don't recognize her description of trees on Pandora: "What we think we know is that there is some kind of electrochemical communication between the roots of the trees, like the synapses between neurons." Sci-fi? Think again.

While Pandora itself is the stuff of legend, tree communication networks, such as the ones described by Grace, are not, as Professor Suzanne Simard of the University of British Columbia discovered in 1997. Trees are, in fact, connected to each another and communicate with one another through their roots. "They shuffle carbon and nitrogen (and water) back and forth according to who needs it," explains Professor Simard. "They interact... trying to help each other survive. Forests are really complex systems... It's a lot like how our brains work."

At the root of all this "connectedness" are tiny, inconspicuous mycorrhizal fungi, which grow on the roots of trees and link up individual root systems to create an underground web. They are, if you like, the stitching that holds it all together. Extraordinary as they are, these organisms make up just a minute part of an enormous, pulsing ecosystem that lies beneath our feet. This ecosystem is one that—as author and ecologist Tony Juniper says in his book *What Has Nature Ever Done For Us?*—"is probably the least appreciated source of human welfare and security" and one which, remarkably, "has taken on the cultural label of 'dirt,' something to be avoided, washed off, or concreted over." In other words—soil.

Above: **A beetle scrambling along a vine at Matassa in the Roussillon.**

Opposite: **Living things don't exist in isolation, and healthy plants are no exception. Developing complex relationships with their surroundings, vines create intricate connection networks above and below ground.**

TEEMING WITH LIFE

Soils are alive, a fact that is too often ignored in modern agriculture (see *Farming Today*, pages 8–11). In fact, "it is estimated that ten grammes (that's about a tablespoonful) of healthy soil from an arable landscape is home

Above: **Compost heaps in Italy at Daniele Piccinin's vineyard (left) and at Johan Reyneke's biodynamic vineyard in South Africa (right). Healthy living soils full of earthworms and other microorganisms are essential for proper plant nutrition.**

to more bacteria than there are people on Earth," Tony Juniper explains. And yet, science knows next to nothing about this soil biology and the complex relationships between it and plants. Indeed, the vast majority of the creatures in a handful of healthy soil remain unidentified to this day.

We do know, however, that the seemingly still world beneath our feet is anything but. There are protozoans eating bacteria, nematodes eating protozoans, others eating fungi, and millions of microarthropods, insects, and worms, to name but a few, all eating and excreting their way through life. And plants are no bystanders. They also get in on the act, exuding food by their roots in order to attract (and feed) the fungi and bacteria that in exchange help feed the plants. When, for example, a vine photosynthesizes, about 30 percent of the fruit of its labors is diverted away from the production of its leaves, grapes, shoots, and roots to be donated to the soil in the form of carbohydrates,

as was explained to me by Swiss natural wine grower Hans-Peter Schmidt. This feeds some five trillion microorganisms (more than 50,000 different species of bug), with whom the vine has established symbiotic relationships. In exchange for this food from above, the micro-creatures supply the vine with mineral nutrients, water, and protection from pathogens from below.

And, as with the trees, these exchange networks also enable communication. "Subterranean communication is not only about mycorrhiza. There is a lot more," explains Hans-Peter. "There are many thousands of different interdependent species of microbes that can even exchange electrons, which is a little like an electric current between plants. And it is these communication highways in the soil that you interrupt when you plow or till."

WHAT PLANTS NEED

As well as facilitating both communication and defence, soil life is also vital for plant nutrition. To grasp this, you have to understand how plants feed. Plants need 24 different nutrients to grow normally and complete their life cycle. (In fact, plants in healthy soil take up more than 60 mineral elements, including iron, molybdenum, zinc, selenium, and even arsenic.) While most of the carbon, hydrogen, and oxygen needed is obtained through the leaves of plants, the rest is only available from the soil. However, plants are unable to absorb these nutrients directly, which means that it is down to the microbes to convert them into forms that can be assimilated by plant roots. Without these all-important bugs, a vine could spend all day extracting trace elements from rocks, but be unable to do anything with them. As Claude and Lydia Bourguignon, two of the most celebrated agronomists in the world, explained to me, "I can show you photos of vines growing in red soils that are rich in iron, but whose leaves are yellow because of chlorosis (iron deficiency). Their roots are literally sitting in the metal, but, because the soils are dead, there are no microbes to process the iron, so the result is a yellow plant in red soil."

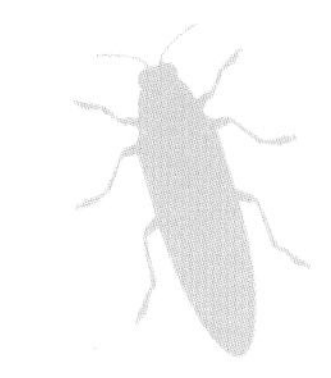

Below: **Living soils on the Styrian vineyard slopes at Weingut Werlitsch in Austria.**

Above: **Living soils are better able to withstand difficult growing conditions, such as drought or heavy rainfall.**

Opposite: **Farm animals in among vines hugely benefit the plants, be they foraging hogs in the hills behind Béziers or large herds of sheep mowing the winter grass at Château La Baronne, a natural wine producer inland from Narbonne. Not only do they help fertilize the soil, but they also increase the biodiversity of microorganisms through their feces and saliva. This contributes to healthier, spongy, living soils—the perfect place for a snooze under the hot African sun at Reyneke in Stellenbosch...**

One of the key factors in the assimilation of nutrients from soil by plants is oxygen, which has to be in ready supply for the microbes to use. This means that the soil needs to be aerated, a job carried out by larger soil fauna such as earthworms which dig tunnels up and down, left and right, so creating whole networks of corridors that are, unfortunately, easily damaged by modern farming techniques.

OTHER BENEFITS

The beneficial relationships don't only occur underground, however. They also take place on the surface of the soil, where a wide diversity of plants and animals makes it harder for pests and diseases to take hold. "The greater the plant diversity, the greater the variety of insects, birds, and reptiles, etc. living in self-regulating competition," explains Hans-Peter Schmidt. Indeed, "where plant diversity is destroyed by monoculture, a negative selection of bacteria, fungi, insects, etc. will occur." In short, everything needs to be in balance and balance is achieved with variety.

What's more, living soils make vines more resilient to extreme weather, which is a valuable asset given today's changing weather patterns. "The permeability of soils treated with herbicides is about 1mm per hour, whereas living soils are about 100mm," explain the Bourguignons. This means that rainwater will infiltrate dead soil a lot more slowly, "keeping the land anaerobic and vines unhappy," as well as contributing significantly to soil erosion. The solution is soil life, which, in addition to ensuring the soil's permeability, also gives it a sponginess that makes it a great ongoing source of water as well. "Organic material in the soil can hold up to twenty times its own weight in water, and thus renders soil more resistant to the effects of drought," says Tony Juniper.

Perhaps most important of all is the fact that soil simply would not be soil without its living parts, since it is the plants and bugs that process the organic matter into humus, one of the fundamental components of soil. No life, no soil.

Natural wine growers know this and so nurture life in the vineyard. They add organic material, such as garden composts, mulches, or cover crops, and reduce soil compaction, creating hospitable environments in which life can thrive. If you take a walk in an organic vineyard, you may witness what looks like a mess—herbs and flowers growing every which way, unruly canes, fruit trees growing in among the vines and, if you are lucky, a few happy cows, sheep, pigs, or geese for good measure. But out of this chaos come balance, beauty, and healthy, living soil.

Soil is clearly one of our greatest assets, an invaluable source of human wealth and well-being, and something that above all needs to be cherished. After all, soil is, as Juniper reminds us, "a highly complex subsystem of our living planet. It is of vital importance and yet it is no more than a thin, fragile skin."

A LIVING GARDEN
WITH HANS-PETER SCHMIDT

Hans-Peter Schmidt runs Mythopia, a 7.5-acre (3-hectare) experimental vineyard in the Swiss Alps, affiliated to the Ithaka Institute for Carbon Intelligence. Alongside grapes, Mythopia also cultivates 5 acres (2 hectares) of fruit, vegetables, and aromatic plants.

"The Valais is a monocultural desert with dead soils. Helicopters spray pesticides and there is virtually no green cover at all among the vines. In fact, for three months of the year, when you drive through the countryside, you have to close your windows because the stench of pesticides and herbicides is so overwhelming. It is actually quite a toxic area. It was a challenge to take over a vineyard here. But, even in the first year, we saw fantastic results. Biodiversity increases really quickly.

After eight years we have birds nesting, dozens of wild animals, including rare green lizards, bees, beetles, wild deer, and rabbits that dart out into the neighboring forest. We have even counted over 60 species of butterfly, which is more than one-third of all butterfly species found in Switzerland.

Butterflies are a particularly good sign of a healthy ecosystem. They are a kind of umbrella species that highlight the health of the overall

environment. We've got striking, polka-dotted *Zygaena ephialtes* (which is a moth, actually), leaf-like Commas, and dozens of rare Iolas Blues that live on the 20 or so bladder senna bushes planted around the vineyard. Iolas Blues are one of the most endangered butterfly species in Switzerland, so we are really privileged to be able to protect them. If you visit neighboring vineyards, though, you won't see more than a couple of species of butterfly at best, whereas here you can spot at least ten species at any time of the year, except winter.

Similarly, you can always find something to eat in the vineyard—salads, strawberries, blackberries, apples, tomatoes... which is just how it should be in a living garden. It is great for creating competition for the vines in the topsoil—forcing their roots deeper—and for providing diverse habitats for both macro- and microbiological life.

We've also introduced our own (less wild) life. Our dwarf Ouessant sheep are perfect partners: they're too small to reach the grapes, but do a great job trimming the green manure and cleaning off the vine trunks, which would otherwise have to be done manually or by machines. Most importantly, though, they increase the soil's microbial biodiversity and organic content, as the sheep's gut bacteria (as well as other decomposing bacteria) are introduced into the vineyard through their feces and saliva. This is essential for fighting soil-borne pathogens, resulting in a healthier soil and vineyard.

We also have 30 free-ranging chickens in among our vines, which is an old Roman tradition. They can help disproportionately with the economics of the vineyard, since our 7 acres could potentially house 500 of them, which would be almost more lucrative than the wine itself!

As biodiversity becomes more complex, so too does the nutrient uptake of the vines, which

Wildlife in and around Mythopia's vines, photographed by Hans-Peter's friend Patrick Rey, including a Marbled White butterfly, hornet moth, and wall lizard. Patrick spent four years watching, following, and recording life across the seasons, as it blossomed among the vines.

themselves become more and more resistant to pathogens. Animals and insects are a key part of a healthy ecosystem.

The benefits of biodiversity are real and easily achieved. When you work at a desk, from an office, writing procedural papers on how to increase biodiversity, actions quickly become overly complicated. But, if you are outside with your feet on the ground, you see just how easily it can be done. Just sticking to a basic principle like 'no vine should be more than 50m away from a tree' will have a radical impact. At ours, we have about 80 trees planted on 2 acres plus the vines, so it is even a little more extreme. Even big vineyards can adapt their setups to this. In Spain, for example, one vineyard I worked with asked me, 'Why on earth should we plant trees when there isn't a single other tree for the next 500km, between here and the sea?!' But they planted them anyway and three years later, they not only noticed the difference, but now swear they'd never do without."

THE VINEYARD: NATURAL FARMING

"An organic farm, properly speaking, is not one that uses certain methods and substances and avoids others; it is a farm whose structure is formed in imitation of the structure of a natural system that has the integrity, the independence, and the benign dependence of an organism."

(WENDELL BERRY, AMERICAN WRITER AND FARMER, IN AN ESSAY ON "THE GIFT OF GOOD LAND")

Above: **A blackbird photographed in Switzerland by Patrick Rey as part of his Mythopia series (see previous pages). The soils of this Alpine vineyard are never tilled or plowed.**

It might be possible to produce natural-like wine (see *The Cellar: Living Wine*, pages 47–50) from non-naturally-farmed grapes because life, particularly microscopic life, is extraordinarily resilient and, even if quashed by chemical sprays, usually manages to pull through in some form or other. However, its complexity, quality, and robustness will suffer, while imbalances in the raw product will usually lead to problems later on in the cellar. If you use fungicides, for example, these will also weaken your yeast populations, making fermentation more difficult and leading to a slippery slope of intervention. Consequently, to be natural, you have to farm cleanly and produce grapes on living soils that are healthy and covered in a rich micro-flora and fauna.

Natural growers use a variety of different farming methods to achieve this, all of which aim to foster plants that are independent of the grower and can fend for themselves. The ideal is to create an environment in which life at large is in balance because, whenever you have an infestation of one particular species, it will take over and problems will ensue. Natural growers, therefore, seek true biodiversity, since plants, bugs, and other wildlife are all allies in the farmer's fight against pests and diseases.

Wine growers will often pick and mix different methods, some of which are explained here.

THE ORGANIC APPROACH

Many of the principles used in organic farming have existed since time immemorial, but an organic consciousness only emerged in the 1940s, thanks to the likes of Sir Albert Howard (1873–1947) and Walter James (1896–1982), who spearheaded the organic movement.

Organic viticulture (like all organic agriculture) aims to eschew man-made, synthetic chemicals in the vineyard. It restricts or prohibits the use of pesticides, herbicides, fungicides, and synthetic fertilizers and, instead, uses plant- and mineral-based products to combat pests and diseases, increase the health of soils, and help build plant immunity and nutrient uptake. (Organic viticulture—which natural producers use—should not to be confused with organic viniculture, or winemaking, as cellar practices in organic and biodynamic accredited wines can be different to those used by their natural wine counterparts—see *Conclusion*: *Certifying Wine*, pages 90–91.)

While organics are pretty big business in most foodstuffs today, wine has been much slower on the uptake. As the biodynamic consultant and wine writer Monty Waldin explains, "[In 1999] I estimated that for the period 1997–1999... just 0.5–0.75 percent of the global vineyard was certified organic or in conversion to organics." Thankfully, today, it's a much prettier picture. Waldin goes on to say, "My best guess is that 5–7 percent of the world vineyard is now organic or in conversion."

There are now dozens of organic certification bodies around the world—including the Soil Association, Nature & Progrès, Ecocert, and Australian Certified Organic—each of which has its own regulations and standards to meet.

THE BIODYNAMIC APPROACH

Biodynamic agriculture is a form of organic cultivation developed by the Austrian anthroposophist Rudolf Steiner (1861–1925) in the 1920s. It is based on traditional practices, whereby polyculture and animal husbandry are at the heart of the farm. Unlike organics, the emphasis of biodynamics is on

Below: **The vines at Les Enfants Sauvages, a biodynamic farm on the Languedoc-Roussillon border in southern France.**

Above: **The moon over a vineyard in Chile. This large satellite orbiting the Earth has a profound effect on our planet.**

Opposite: **A wasp's nest among the vines at Strohmeier, in Austria, which is one of the country's most avant-garde natural wine growers.**

prevention rather than treatment, as well as on encouraging the self-sufficiency of the farm unit. Natural preparations based on plants (yarrow, chamomile, nettle, oak bark, dandelion, valerian, horsetail, etc.), minerals (quartz), and manures are all used to stimulate microbial life, boost the immune systems of plants, and improve soil fertility.

This holistic approach treats the farm not in isolation, but as part of a landmass that is part of a planet that is part of a huge solar system, where large bodies of mass exert considerable forces (gravitational, light, etc.) on one another. Life on Earth is fundamentally affected by these large external factors—and biodynamics takes this into account.

While people sometimes struggle with this astronomical approach to farming, some of it is really just common sense. As astronomer Dr. Parag Mahajani once told me, as I stared through the eyepiece of an enormous telescope, "People don't realize how bright the moon can be. When it's a full moon, plants grow more."

Similarly, consider the tides and the effect of the moon's gravity on our oceans. It doesn't take much to realize that plants—which are mostly made from water—are greatly affected, too. As Dr. Mahajani also told me, "Tides have one of the most profound effects on Earth. Gravitational pull is on everything—on gases in the air, on land, and in water. Everything is moving up and down—all the buildings, roads, walls, concrete, everything experiences the tide. But, since the bonding of molecules in solids is stronger than those in liquids or gases, it's just less obvious." This awareness informs biodynamic farmers' choices, including, for example, when to prune their vines or bottle their wines. For some fascinating practical illustrations of biodynamics, take a look at the works of Maria Thun (1922–2012)—see *Further Exploring & Reading*, pages 217–18.

OTHER WAYS OF FARMING NATURALLY

Two personal favorites of mine include the teachings of Masanobu Fukuoka (1913–2008) and what seems to me to be his Anglo-Saxon counterpart: permaculture or "permanent sustainable agriculture."

Fukuoka was a Japanese farmer-philosopher celebrated for his so-called "hands-off" approach to farming, which had amazing results. As he explained in his book, *The One-Straw Revolution* (published in 1975), Fukuoka managed to achieve similar rice yields, with his no-tilling, no-irrigation, no-herbicide methodology, as his conventional counterparts toiling day in and day out in neighboring paddy fields.

Permaculture, on the other hand, was a word originally coined in the 1970s by Australians Bill Mollison and David Holmgren. As Mark Garrett, a permaculturist friend of mine, once explained, "It's 'a way of seeing' or 'a way of looking at' agriculture, so that you think about processes and design systems in such a way that what you set up is self-sustaining and self-sufficient.

Below: **Chickens at Le Petit Domaine de Gimios. Animal husbandry is a fundamental part of holistic farming.**

Below right: **Daniele Piccinin (whose estate carries his name) uses his own plant concoctions to treat his vines (see pages 76—77).**

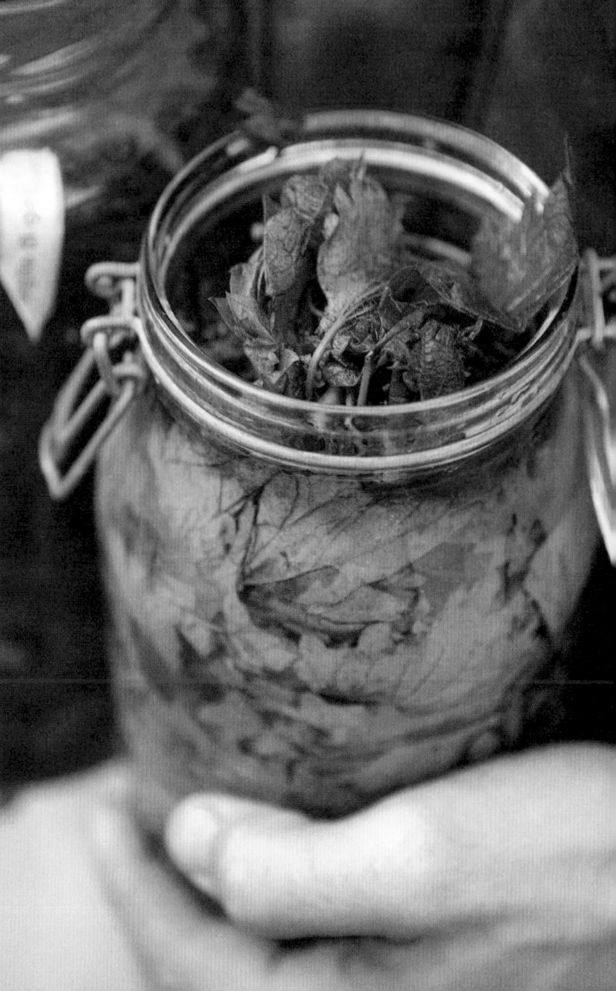

There is *no one single* permaculture: different contexts, and different scenarios, necessarily mean different permacultures. You could have one that uses organic principles, one that uses biodynamics, or one that doesn't label itself as any one thing in particular. Permaculture encapsulates an idea shared by many cultures around the world: that we should farm in such a way that we enrich our environments both for ourselves and for all life that depends on that place, including future generations."

In the end, whether you call it organic, biodynamic, or permaculture, it's not the label that counts, but the motivation. In my experience, while any attempt at "clean" farming will have a positive environmental impact, "turning green" for marketing reasons never results in very inspiring farms. Your heart has to be in it. Converting to clean farming can be tough, particularly at first, so something else has to drive it. You have to do it because you know that, in the long term, there is no other way, not because it will get you new customers.

Below: **Frank Cornelissen, seen here tending the vines in his magma-producing vineyard, Barbabecchi, on the slopes of Mount Etna in Sicily, uses Fukuoka-inspired low-intervention farming practices.**

DRY-FARMING WITH PHILLIP HART & MARY MORWOOD HART

"When we started, we decided to go super high-tech. I thought, I can be on my computer up in Orange County and can turn on the sprinkler system... We had a James Bond consultant. He was from this company of newest, hottest guys. They wanted to put probes in the ground so that we could know the moisture content remotely. We were totally into the idea.

Yet, we travel the world and we know the reality of old vineyards that are irrigated and we know that it's possible to do otherwise. So, we said to this consulting person, 'Why don't we dry-farm?'

'No, you can't do it.' He was right out of college and none of the universities round here—UC Davis, Berkeley, CalPoli, Sonoma State—teach alternative methods, as they're perceived as not being economically viable.

So, while we loved the idea of dry-farmed bush vines, we were going to go conventional high-tech anyway.

That is until one day when we happened to pull into a nearby cellar we'd never seen before. The woman behind the bar was slightly ripped. I'm not kidding, she poured us the most giant glass of red wine, and Phillip and I were like 'Oh.' What are we going to do if we don't like this wine? We took a sip, it was a sangiovese-cab blend and, my gosh, we loved it.

AmByth—or "Forever" in Welsh—Estate is a dry-farmed, 20-acre (8-hectare) organic vineyard and cellar in Paso Robles, California. The husband-and-wife team cultivate 11 varieties of grape. They also keep bees, chickens, and cows, and grow olives.

'Where's this from? How's it farmed?'
'It's from right here—it's dry-farmed.'
'Who planted the vineyard?'
'My husband did.'

We met him the next day. He's an old-time vineyard guy who's always dry-farmed and he said to us, 'Look at them weeds. If them weeds can grow, vines can grow.'

So, we fired the super high-tech company—and didn't look back.

In Paso Robles, we have a huge water shortage. My numbers may be off a little but, basically, our water table has dropped 100ft in the last 10 years, which is directly caused by vineyards. We've been told that there are 20,000 more acres going in in the next few years, so what is going to happen to our water table? Environmentally, it is completely unsustainable. Non-irrigated land is being transformed into majorly irrigated vineyards. And the rain isn't replenishing what is being consumed. So, it doesn't take a genius to figure out the problem. People's wells are going dry.

The sad thing is the huge blocks of new plantings being bought by outsiders. It's not Mary and Phillip who've bought 200 acres next door to range cows. It's somebody from Los Angeles or China, so the water drying up doesn't mean much to them. They dry up their neighbors and move on. It's just a financial deal.

'If we buy 800 acres and plant 600 acres, what crop will we get in two years?'

'How quickly will we get a return on our investment?'

And, yes, if you do the numbers, they can get it all back in about four years. After that they don't care. It's all gravy. And if it doesn't work, they walk away. With dry-farming, the return is much slower but the result, as our name suggests, is forever.

We're one of the driest farming areas in California. We get tonnes less water than Napa. We even get half the water they get just over there—on the other side of the 101. So, if we can do it, they should all be dry-farmed.

Our basic belief is that a vine is a weed. Vines love to grow. They're survivors. They are like the cockroach of the plant kingdom, and isn't that wonderful!"

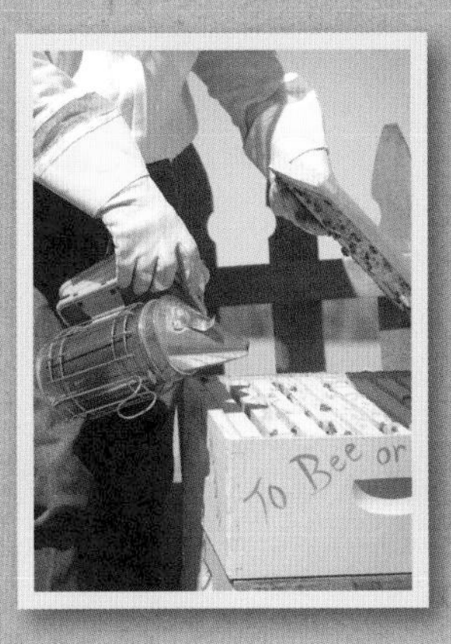

Above: **Unlike the vast majority of vineyards in California, AmByth's bush vines are dry-farmed.**

Left: **Mary and Phillip tend their bees themselves. "Our honey is thick, deep, dark, and rich because our bees eat their own honey. We get about 40lbs of honey per hive, and we leave at least half per hive for them."**

THE VINEYARD: UNDERSTANDING TERROIR

Opposite: **Factors such as soil composition, climate, exposure, and altitude all contribute to *terroir*.**

To understand why natural wine is so special, we have to take a step back and look at what is known as *terroir*, which "good" natural wines invariably express. Simply put, *terroir* is a French term (derived from the French word for "earth") that has come to refer to "a sense of place," a unique, irreproducible combination of factors (plant, animal, climate, geology, soil, and topography, etc.) in a particular year.

It is a word that describes an agricultural context which can also be applied to olive oil, cider, butter, cheese, yogurt, and the like. "It is a glorious concept that came about when people living in a particular place noticed that plants, or animals, raised in that location expressed flavors that were not possible elsewhere. It was a powerful guarantee for consumers," explains Nicolas Joly, a biodynamic producer from the Loire, in France, and founder of the growers' association *La Renaissance des Appellations*.

Humans may also be a part of this context—but they are only ever part of it. If they dominate, then the expression of the place dwindles. Anselme Selosse, one of Champagne's most iconic producers, explains this distinction well: "As a young winemaker, it was out of the question that I be subservient to nature. I was determined to be boss. I dominated the vines and wines entirely. And, although I was making wine exactly as I had wanted to, none of the results captured my interest. That is, until I realized that my way of being was totally unconducive to the creation of great art, since the originality, or singularity, of a place, which I so fervently sought, was in fact *entirely* dependent on my giving it the freedom to express itself."

Different years produce different growing conditions, which, in turn, affect all the life forms living in that place, each making the most of the resources on offer and each inextricably linked to the rest through symbiosis, dependence, the food chain, or simply because they happen to be in the same place at the same time. The result is an incredibly intricate web of inputs that is infinitely

"Nowadays is the first time in the human story that we are able to make wine without terroir, just with chemicals."
(CLAUDE BOURGUIGNON, AN AGRONOMIST FROM BURGUNDY, FRANCE)

more complex than anything that man can create. Nature, in all her profound subtleties, can always do it better.

As Jean-François Chêne, a natural grower in the Loire, in France, explains, "Each year we reproduce the same gestures, but they're never quite the same. There are always little differences because of the year and that's what is interesting." It is, however, possible to iron out these vintage variations through interventions in the vineyard (such as using weed-killers or even irrigation) or in the cellar (see *The Cellar: Processing & Additives*, pages 54–55). In fact, many of today's wines iron them out completely in the name of brand consistency.

Wine is an agricultural product created by living organisms in a particular place at a particular moment. It is the product of *life* forms, the sum of which is *terroir*. And, without them, *terroir* cannot be expressed.

As grower Anne-Marie Lavaysse from Le Petit Domaine de Gimios, a vineyard in the Languedoc, in southern France, explains, "Natural wine is all about what Nature gives me. It is simply the result of what the vineyard gives me each year." Since it is a drink that nurtures and sustains life across the board, it is literally *full* of it—from the vineyard, to the cellar, to the bottle, and then in the glass.

As Jean-François Chêne succinctly puts it, "The most important thing for me is to respect the living above all else."

Above: **One of Etienne and Claude Courtois' biodiverse vineyards at Les Cailloux du Paradis in Sologne, France, surrounded by native woodland.**

Opposite: **Autumn at La Coulée de Serrant, Nicolas Joly's biodynamic estate in the Loire, France.**

SEASONALITY & BIRCH WATER

WITH NICOLAS JOLY

"Every planet is linked to a tree type. Birches, for example, are Venusian. Stand next to a birch and it's not like standing next to an oak. It's not rigid. It's not massive. It doesn't shout, 'I'm here' or climb over others. Instead, the birch is extremely supple, ever-changing. If you look at the shape of a birch, it almost looks unfinished. It is not final like the cypress, for example, that explodes in the shape of a candle flame. The birch is not demanding and grows easily just about anywhere.

Someone once told me, 'If you want to understand what Venusian means, imagine that you have people in your home and everyone is speaking animatedly. All of a sudden a quiet person discreetly enters the room and places a cup of tea in front of each guest, saying, "I thought you might like some refreshment."' That's a Venusian approach. Softly. Sensitively. Simply. It's feminine energy.

When you collect birch sap, you're collecting the essence of spring. The reawakening of the new beginning, when everything is bursting into life. Drinking it, therefore, is a real act of rejuvenation, which is why people like Weleda use it in so many therapies. It is quite a common tree, so anyone can have a go at extracting the water for themselves. Just remember to do it with care and an awareness that you are nourishing yourself with the work of an elemental being.

Nicolas Joly is one of France's foremost authorities on biodynamics and founder of the wine growers' association La Renaissance des Appellations (see "Where and When: Grower Associations," pages 120–21). Author of multiple books, he is also a leading advocate of spontaneous fermentation and indigenous yeast. He owns the 900-year-old vineyard La Coulée de Serrant, in the Loire.

HOW AND WHEN TO HARVEST

Birch water is the spring of the leaf. It embodies the very beginning of the cycle, the moment before bud burst, before the sap becomes leaf. It's the moment when, if you feel carefully, you can sense that nature is on the move, but nothing is quite visible yet. That's your window of opportunity. You have 10 to 20 days (ideally when the moon is rising) during which the birch sucks up a massive amount of water from the ground, drawing it upward in its sap toward the burgeoning leaf buds that are not yet visible. This suction creates extraordinary pressure, which is when you can harvest. Dates vary enormously. At ours, it is usually between February 20th and March 4th.

You'll need a little wooden hand drill, about 5mm wide, a large empty water bottle, and a clear syphon hose, like the sort you'd find on a lawnmower's carburetor. The hose needs to be the same diameter as the drill, so you're better off buying the hose first and the drill second.

Pick your spot, and pierce the bark a maximum of 2cm deep. You'll know soon enough if you've picked the right moment. The tree will be ready to burst with the pressure of the water, so your little hole will drip right away.

Stick one end of your hose in the hole and the other in the empty bottle, strapped tightly to the tree. When the birch is in full swing, you'll have to stop by once a day to empty the bottle. I collect up to 1.5 liters a day.

What you have to bear in mind is the importance of respecting the tree. If the first hole you drill doesn't seem to give, don't drill another. You might be too early in the season, so come back regularly to check. The sap was destined to become leaves and it is an exhausting process for the tree. If you take a little for your own consumption, that is fine, but you can't push it without causing damage. One hole per tree, never more. You're better off not making any holes unless the process can be supervised from start to finish as, once it is flowing, you can't block the hole. It will flow until it has amassed enough water for its leaf-creation process, which can take three weeks. So, once you start, you can't stop. You will have to harvest daily, much like milking a cow.

When the flow does stop and the bark is dry and free of humidity, which happens naturally at the end of the process, remove the hose and the tree will heal itself. However, as a symbol of gratitude, you can help the tree by using a little pine tar (also called Stockholm Tar) to plug the hole. You only need a tiny amount, about the size of the tip of a ballpoint pen, so don't buy the synthetic rubbish, which will harm the tree. Once done, thank the tree and, remember, you're dealing with a living being.

I collect about 30 liters per year. It keeps for months, especially in the fridge. Drink a glass a day on an empty stomach, first thing each morning, at the dawn of your daily spring."

Below: **A Coulée de Serrant vineyard in winter.**

THE CELLAR: LIVING WINE

"Looking at it under a microscope,
natural wine looks like a small universe."
(GILLES VERGÉ, A NATURAL WINE GROWER IN BURGUNDY, FRANCE)

The French term *vivant* (meaning "alive") often crops up in relation to natural wine, as do phrases such as wines with "soul," "personality," or "emotion" that attribute living, human-like traits to what most people consider to be an inanimate drink.

Having decided to take a closer look at this "life," I enlisted the help of Laurence, a scientist friend and academic, who also had access to microscopes. I gave her two bottles of Sancerre: a conventional, large-volume, own-label store bottle, produced in the tens of thousands, and *Auksinis*, a wine by Sébastien Riffault, of which fewer than 3,000 bottles are produced annually. *Auksinis* is a natural wine in the most complete sense—there's nothing added and nothing removed.

A couple of months later, Laurence emailed me photos of the wines she'd taken under the microscope. The contrast was astonishing (see right). *Auksinis* was full of yeasts—some were dead, many were alive—while the large grocery store wine was seemingly lifeless. Laurence had even managed to culture what she thought were lactic acid bacteria filtered out of Riffault's wine. *Auksinis* was full of microorganisms and yet, contrary to what most people in our disinfected Western world would think, it was both perfectly stable and absolutely delicious. It was a 2009, with a nervous acidity, a touch of smokiness, and notes of acacia, honey, and linden (lime), which was aromatically as clean as a whistle. Not an off-aroma in sight.

On the face of it, there was little to distinguish one bottle from the other—both were Sancerres and both were sold as such in the United Kingdom. But, inside the bottle, it was chalk and cheese. Certainly, the taste profiles of the wines were totally different, but the lack of resemblance went far beyond a subjective "I like that" or "I don't." They differed fundamentally in terms of microbiology. *Auksinis* was full of it; the store wine was not. So, as Laurence

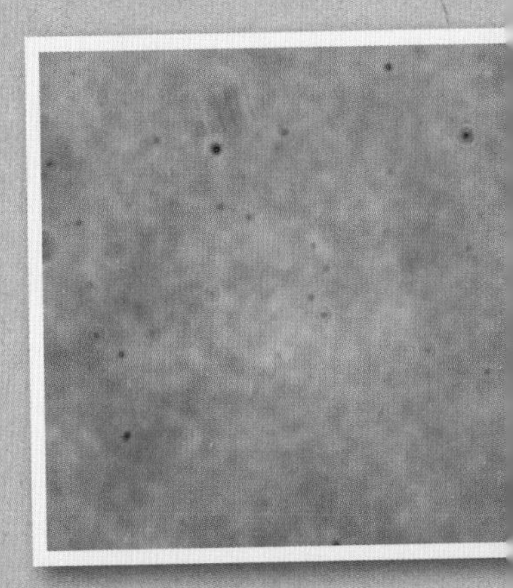

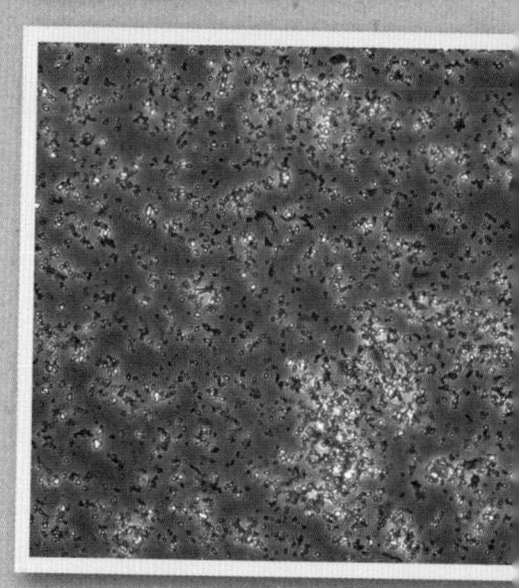

Above: **Under the microscope: a large-store Sancerre (top) and Riffault's natural Sancerre, *Auksinis* (bottom).**

PAS COMME LES AUTRES
CAVE A MANGER
VINS VIVANTS
BEZIERS
Tél. 04 67 48 53 05

sat down that evening to a glass of Riffault, after putting a few petri dishes under the microscope, she was not only literally drinking a live product, but also getting an actual living taste of Sancerre.

THE SCIENCE

Three recent-ish scientific studies are a further testament to this "living-ness" and its ability to continue for decades, if not centuries. Firstly, in 2007, the American Journal of Enology and Viticulture published research into the "Survival of Wine Microorganisms in the Bottle During Storage." Having explored a selection of Bordeaux wines across various vintages, with the oldest dating back to 1929, the team found that most of the older bottled wines contained elevated yeast populations. In fact, a Pessac-Léognan bottled in 1949 contained 4 million cfu/ml (meaning colony-forming units per milliliter), which, according to the authors, amounted to between 400 and 4,000 times as many as the average microbial population found in many pre-bottled wines today. They also found that 40 percent of the bottles contained lactic acid bacteria.

Then, in June 2008, Professor Dr. Jürg Gafner, of the Agroscope Wädenswil Research Institute in Switzerland, explored the microbiology of a selection of Räuschling whites, the oldest dating from 1895. To the surprise of many, he isolated six different, *living*, dormant yeast strains across the various vintages, including three from the oldest.

Finally, and perhaps most remarkably of all, a Jurassien *vin jaune* ("yellow wine") from 1774 was tasted some 220 years later by a group of local experts, who described it as "nutty with notes of curry, cinnamon, apricots, and beeswax with an exceptionally long finish." According to microbiologist Jacques Levaux, who later tested it at the laboratory, both bacteria and yeast were found and, although dormant, they were still very much alive.

The bacteria, in particular, may hold an interesting secret, as Dr. Karin Mandl from the biochemistry division of the HBLA und Bundesamt für Wein- und Obstbau, in Vienna, explained to me. Although still at the very beginning of her research, Karin is trying to cultivate bacteria found in various wines in the hope of isolating a strain responsible for wine's ageability. Gilles Vergé, a natural wine grower in Burgundy, is also convinced: "Without bacteria, wine cannot age. It's they that keep freshness going even in very old wines. They can continue to evolve for decades, if not centuries," he says. "They don't need much to survive, just the trace sugars that remain after fermentation."

THE BENEFITS

So life is key, not just for growing grapes or for fermentation, but also perhaps for wine's potential to grow old gracefully. This is not to say that only natural wines contain life—after all, all wines are made with yeasts and bacteria

Above: ***Élevage* is fundamental to living wines, which find their stability with time and age. Natural wines can be very long-lived and bacteria may well be at the root of this ageability.**

Opposite: **Natural wine producers use indigenous yeasts to create living wines, which has a positive impact on the taste of the wine. As Nathalie Dallemagne, the Technical Consultant for Viticulture and Winemaking, CAB (*Coordination Agro-Biologique des Pays de la Loire*) explains, "If you look at fermenting must under microscope, you can immediately see if commercial yeasts have been used. They are usually bigger than wild ones and the cells are identical since they are all the same strain."**

Above: **The Radikon cellar, where wines mature for years in barrel before release.**

Opposite: **The late Stanko Radikon and his son Saša (pictured) have been making traditional, sulfite-free, natural wines for decades.**

(whether indigenous or not), so, at some point in the process, all wines must have been alive. It is just that natural wine is a lot more "alive" than the rest. Many conventional wines will include microbe populations, but how many depends on multiple factors, from agricultural practices through to the processing and additives used in the cellar. Extreme filtration, for example, may well have been the main reason why there wasn't any yeast in sight in the store example photographed by Laurence.

By impacting on the microflora, these artificial interventions also seem to have an impact on taste. "Chemical wines are like a flat line," says Saša Radikon, whose family has been making sulfite-free wines in the Collio, in eastern Italy, since 1995. "How long the line is depends on how good the enologist is but, fundamentally, it's a flat line that suddenly comes to an end. Natural wine, on the other hand, is like a giant wave. Sometimes it shows well, others less, and, like all living things, it will eventually die, but it might be tomorrow or in 20 years' time." This, Saša explains, is intrinsically linked to the life inside the liquid, which is noticeable throughout the year. "Our cellar is not temperature-controlled, so it moves with the weather. In winter, when things are slow outside, the wine is slow too. Then, in spring, when life picks up again, so too do the wines. They have more flavors and they taste different. Then back to fall, winter, and to sleep again. The wine is definitely alive."

Proper living wine is, literally, a kaleidoscope of taste. Try it today, and it will show certain aromas; taste it tomorrow, and it will be different. These are wines with an ever-changing, extraordinarily complex aromatic profile that functions a little like a child's mobile—different parts rotate individually, showing themselves differently at different moments, so that the whole is never the same twice. Sometimes open, sometimes closed. Sometimes giving, sometimes not. It is almost as though the microbiology needs time to wake up and react, or simply choose to sulk in the corner.

Just as we are beginning to talk about the "second genome" or the fact that, as humans, we are so much more, biologically speaking, than just an "I"—as bearers of genetic information, we're more than 99 percent other stuff as well, according to Michael Pollan of the *New York Times*—so, too, wine is more than a simplistic combination of organoleptic compounds, alcohol, and water. It contains "other stuff" as well. And, like *our* other stuff, that of wine also lives, protects, defends, conquers, grows, reproduces, sleeps, ages, and dies. This is fundamental to what makes wine wine, rather than a simple, sterile, manufactured alcoholic drink.

MEDICINAL VINEYARD PLANTS
WITH ANNE-MARIE LAVAYSSE

"I've never liked the sort of medicines that doctors prescribe. Instead, I use wild plants to treat myself, my children, and my animals. So, it seemed like a logical extension to do the same with my vines. What better way to keep a vine healthy and happy than by using another plant?

I let wild grasses grow all over my vineyard, so my vines are surrounded by the southern French garrigue. All sorts of plants make it up and each has a strong, particular smell. It dawned on me that this was the answer. The plants in the garrigue were my vines' neighbors. They lived and worked alongside each other. They had the same experiences and, yet, the garrigue wasn't sick. I already knew a few of its plants; more specifically, I knew of two or three that were great for cleansing and detox. I figured that it was important for the vines' sap to flow and for them to be able to get rid of toxins in their systems, so I went with my intuition. Once I started *really* looking, it was almost like the right plants spoke to me.

I would stir and macerate the plants in the sun, and apply the concoction to my vines. It was extraordinary. The vines were beautiful, wonderful—no oidium whatsoever. I've been doing it now for 10 years and we're still going strong.

Anne-Marie Lavaysse and her son Pierre own Le Petit Domaine de Gimios, a remote 12-acre (5-hectare) biodynamic vineyard in the Languedoc's Saint-Jean de Minervois, an area famous for muscat.

What plants do I use? Well, it depends on what I am trying to do. There are antiseptic and antibiotic plants, which I use to fight infections, or to help reduce a fever, and others that cleanse or regulate. They can all be used on vines as well as humans.

Sage (*Salvia officinalis*), for example, is hugely purifying for the liver. It's great drunk as a tea, but also good for the vines, since the same properties that cleanse the liver also detox plants. Sage is also an antiseptic, so it's good for helping to eliminate fungi that want to install themselves on the plant.

St. John's wort (*Hypericum perforatum*) is another great healing plant, which grows wild in among the vines. It has bright, sun-yellow flowers that are very beautiful. I dry the very top of the flowering part of the plant. This can then be used in a tisane, which is soothing and calming. It helps your muscles relax and is great for getting to sleep. It acts on the neurosensory parts of the body, so is very effective as an antidepressant, as well as a painkiller for both humans and animals. Alternatively, you can leave the flowers in oil for three weeks in the sun and then apply this locally to heal burns or help with muscular aches and pains.

Yarrow (*Achillea millefolium*) is another cleansing plant, which is great for women. If it's that time of the month and I have stomach ache, I make myself a yarrow tisane using the flowers. You can use a little of the leaves as well if you like. It's very effective. It is soothing and helps regulate the system really well. I use it on the vines as well when necessary. Yarrow contains natural sulfur, so its anti-cryptogamic properties are useful for protecting against oidium. Yarrow also helps heal internal tissues, so it works well on the vines' sap 'veins,' which can become blocked if the vine is unwell or has received incorrect treatments.

Right: **Anne-Marie uses medicinal plants such as this *Cistus* in order to treat her vines. It goes into her tisane mix because of its anti-fungal properties.**

Then there's **boxwood** (*Buxus sempervirens*). This plant is toxic, so you have to use it carefully and properly. Boxwood flowers are antibiotic and its leaves are fiercely cleansing. If you have a fever, for example, the leaves will make you sweat, which helps get the bug out of your system. I pick boxwood when it's in season and then stock it at home for use when needed. I boil the leaves for five minutes, filter the water, and then drink it. If you've caught a bad cold or have a high fever or something that is making you feel really unwell, you continue boiling and drinking for two days. It's very effective. You can even use it externally as an antiseptic. It heals wounds quickly."

Right: **Anne-Marie also collects wild herbs and plants for healing and feeding her family, as is the case with this wild fennel (*Foeniculum*), which she uses in cooking.**

THE CELLAR: PROCESSING & ADDITIVES

"So many poisons are employed to force wine to suit our taste —and we are surprised that it is not wholesome!"
(PLINY THE ELDER, *NATURAL HISTORY*, BOOK XIV, 130)

Above: **Sorting tables, used to check the quality of the grapes, at Donkey & Goat in California.**

Most people regard wine as an artisan product that is made from grapes using simple equipment such as a press, a pump, oak barrels or vats, and a bottling facility. What many people don't realize is that it is often a lot more complicated than that. In fact, apart from sulfites, eggs, and milk, most additives, processing aids, and equipment are used behind closed doors because of shortcomings in wine-bottle-labeling laws. "In the US, you can even use anti-foaming agents," says Californian natural wine producer Tony Coturri. "Like that when you fill up your tank, there's no waiting for the foam to drop. If it were chickens or any other food that anti-foaming agent was being added to, they'd say, 'Hang on, what are you doing?!' The FDA [Food and Drug Administration] would close you down."

It is perhaps because of this that winemakers are often reluctant to discuss exactly what is used when, even if what they are doing is completely legal. The result is that the industry as a whole is mostly cloaked in secrecy. You'd be amazed how often I find myself tasting with wine reps who know surprisingly little about the wines they're selling, beyond which grapes were used, whether the wine saw any oak, and for how long it was matured.

High-tech gadgetry exists that can, for example, adjust the alcohol content or perhaps micro-oxygenate or sterile-filter the grape juice. Cryoextraction, for example, is used to freeze grapes, solidifying their water content so that it stays behind during pressing. This is commonly used to mimic the naturally concentrating effect of noble rot—or *Botrytis cinerea*, a form of gray mold—in classic, sweet-wine-making regions such as Sauternes. Other invasive equipment includes the reverse osmosis machine, which is capable of separating out wine components and removing them if desired. You can, for instance, zap out water (if it has rained a lot) or alcohol, remove the taint of bushfire smoke, or eliminate yeast strains that produce "unfavorable" flavors like *Brettanomyces* (see page 78).

As far as additives and processing aids go, the number and specifics vary according to the country—over 50 groups of products, including the likes of haemoglobin, are allowed in Mercosur (Argentina, Brazil, Paraguay, and Uruquay), for example, while over 70 groups are permissible in Australia, Japan, the European Union, and the United States. These range from simple things like water, sugar, and tartaric acid to more obscure powdered tannins, gelatin, phosphates, polyvinylpolypyrrolidone (PVPP), dimethyl dicarbonate, acetaldehyde, and hydrogen peroxide. Animal derivatives are also prevalent, including albumen and lysozyme (from eggs), casein (from milk), trypsin (extracted from the pancreas of pigs or cattle), and isinglass (an extract from the dried swim bladders of fish).

Such manipulations—be they heavy processing, additives, or aids—are usually intended to save time and increase the producer's control over the winemaking process, especially if the winery's operations are on a large scale. Commercial realities, such as the need to release wine within a few months of harvest for cash-flow purposes, mean these interventions are sometimes confusingly referred to as "necessary." Wine is one of those rare drinks made from a primary material—grapes—that naturally contains everything the wine needs to exist. This means that anything else should be regarded as an extra. As Gilles Vergé, a natural grower in Burgundy, France, who keeps his wines in barrel for four to five years before bottling, told me in the fall (autumn) of 2013, "Look at what's on sale now, most people are selling their 2012s. Before, people waited two to three years minimum before selling bottles. They waited for wine to clarify itself naturally, but now they accelerate the process. Soon Beaujolais Nouveau from 2013 will be released, whereas my 2013s still look like mud! It's not possible to do that naturally."

This is one of the factors that sets natural wine producers apart. They are not creating a product simply to supply a demand. They think of their wines as children rather than commodities and are, in general, not looking for the easiest solution, but one that is most enriching for themselves, their plants, and their land. They produce wines of *terroir*—uncompromisingly and come what may. They eschew gadgets, processing aids, and additives, as these make the wine less "true." As Anselme Selosse, a cult Champagne producer, says, "The vine is where it all happens, where everything is captured. It's here that you reach 100 percent of your potential. You can't then add to this during vinification. You can destroy or hide elements, but you can't get extra credit in the cellar."

Above: **Foot treading is the most traditional way of crushing grapes. It is extremely gentle on the berries and is a technique still in use today.**

THE CELLAR: FERMENTATION

> "Fermentation is nature's way of recycling nutrients back into the soil. Microorganisms sit on the berries, waiting for access. When these finally drop, the berry's protective skin cracks and in go the microbes to break it down, returning its components to the soil so that they can be taken up by plants once again."
>
> **(HANS-PETER SCHMIDT, FROM THE MYTHOPIA VINEYARD, SWITZERLAND)**

Fermentation happens when yeasts, bacteria, and other microorganisms break down complex organic compounds (i.e. plants, animals, and other stuff made of carbon) into smaller chemical components. It is a process of decomposition and, in the case of wine, is the single most important part of vinification. It is here that sweet grape juice is transformed into an alcoholic drink and when many of the aromas that make wine so interesting are born. Left to its own devices, fermentation is usually a two-stage process in which alcoholic fermentation (by yeasts) is followed by malolactic fermentation or malo (by bacteria).

Miraculously, these agents of change occur all around us. For example, there are around one million bacterial cells in a milliliter of fresh water and tens of millions of yeast cells in a milliliter of healthy, organic grape juice. Little wonder, then, that "by the time we are fully grown, we have a 3-pound organ of 'others' inside our guts," says Carl Zimmer, science writer for the *New York Times*. Some are benign, a few are pathogens, and many are essential to our well-being.

Having fermented foods at home, and produced several thousand bottles of wine, I have an immense respect for these invisible soldiers that crack on with their task spontaneously, producing great transformations and delicious, *live* flavors. Leave flour and water on your kitchen counter, for example, and a culture for sourdough bread will start up. Leave grape juice in a bucket and it will become wine, or perhaps vinegar, depending on which microorganisms take hold. In fact, both yeasts and bacteria play a key role in making some of the most exciting foods we know, including cheese, salami, beer, cider, and, of course, wine. Certainly, not all yeasts or bacteria are always desirable, but, if you strengthen those that are good, they usually have a fighting chance of conquering and guarding the space they come to occupy.

Above: **Healthy grapes from a healthy vineyard spontaneously ferment if left to their own devices.**

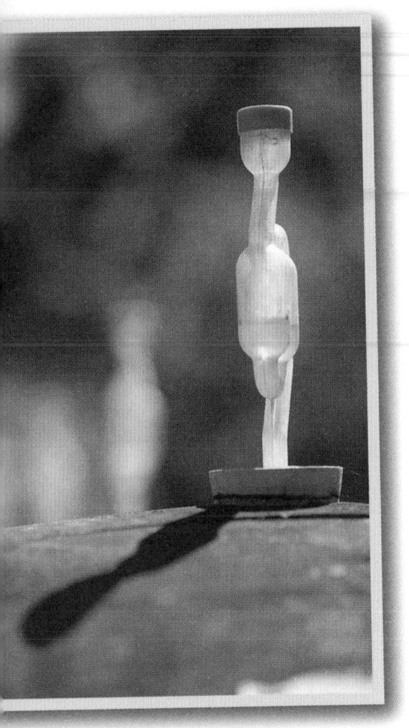

Above: **Airlocks at Tony Coturri's vineyard in California. Inserted into the tops of barrels, they allow carbon dioxide to be released during fermentation.**

Opposite (top left): **Spontaneous fermentation.**

Opposite (top right): **Ridding the barrel of grape solids after racking off freshly fermented wine.**

Opposite (bottom left and right): **Tasting tank samples to monitor the wine's development.**

WHAT YEASTS DO

Yeasts are invisible fungi capable of exponential reproduction when in the right place at the right time. More specifically in the case of grapes, yeasts are found everywhere, from the soil to the vines to the cellar. Their job is to consume sugars in the grape juice, releasing alcohol as a by-product, alongside carbon dioxide and complex flavor compounds. They are the crux of natural wine. They are part of the *terroir*, along with soil, grape, climate, topography, etc. Yeast populations vary from year to year as a result of environmental factors, so contributing to what is known as vintage variation (see *The Vineyard: Understanding Terroir*, pages 40–43). These diverse populations kick in at different stages of the fermentation process in a domino effect that sees a new yeast pick up where an older one leaves off. In particular, the sugar-loving *Saccharomyces cerevisiae* yeast, so essential in baking and brewing, is quick to take over from other yeast strains and fundamental to the production of wine.

Yeasts find strength in numbers. An abundance of yeast is necessary to carry out an effective natural ferment, while using a diversity of strains results in layers of different flavors in the end wine. As Pierre Overnoy, from the Jura, a region in the east of France, explains, "When the official 1996 harvest day was declared, we measured our yeast population. It was 5 million cells/ml (equivalent to a drop of juice). Lots of our neighbors started harvesting, but we held off for a week, until the population had reached 25 million/ml." For Pierre, who does not add sulfites at any stage, the health and size of the yeast population were paramount. "In order to avoid problems during fermentation and achieve great complexity of flavor, you need as many yeast as possible."

Natural fermentations can take longer than conventional ones because the growers are working with unpredictable wildlife. The fermentation can last from as little as a couple of weeks to several months or even years.

What conventional producers do differently: These producers often eliminate indigenous yeasts (using heat, sulfur dioxide, and filtration, etc.) to inoculate lab-bred strains that have been tested to reduce risk, encourage specific flavors, and speed up production. "It's interesting to see the parallels between the descriptions used by yeast manufacturers and those used by the wine trade when describing *terroir*," says Nicolas Joly, who advocates spontaneous fermentation. "Consumers should be told that the aromas in wine are often added in the cellar." Indeed, brochures for commercial yeasts certainly make for interesting reading:

BM45—Italian isolate recommended for Sangiovese. Contributes higher acidity, low astringency, and… great mouthfeel… Brings out aromas described as fruit jams, rose petals, and cherry liqueur, with notes of sweet spices, licorice, and cedar… perfect for creating traditional Italian wine styles.

CY3079—"Classic" white Burgundy: floral notes, fresh butter, toasted bread, honey, hazelnut, almond, and pineapple… rich and full mouthfeel.

Peter

LA FERME DES SEPT LUNES
SAINT-JOSEPH 2011
CHEMIN FAISANT
SYRAH
2012
LA FERME DES SEPT LUNES
glou
Vin de France
Jean Delobre
PAYSAN VIGNERON - 07340 BOGY
PRODUIT DE FRANCE
VIN BIOLOGIQUE
AB
AGRICULTURE BIOLOGIQUE
L. GR12
12,5% vol
750 ml

WHAT BACTERIA DO

Millions of bacteria work alongside the yeasts. Like yeasts, they cover berries and line cellar walls. One key beneficial strain called lactic acid bacteria (or LAB)—think of the probiotics in live yogurts—plays a fundamental role in creating wine. It carries out a secondary fermentation called malolactic fermentation (or malo) during which malic acid, which occurs naturally in grape juice, is converted into the softer lactic acid, so changing the texture and flavor. Although not strictly speaking a fermentation, it is referred to as such because of the bubbling effect of the CO_2 that is released during the transformation.

Natural wines mostly go through malo because, left to their own devices, the bacteria will take over once the yeasts have finished the alcoholic fermentation (although sometimes this can happen before the end of the alcoholic fermentation, which is risky as it can lead to volatile acidity). Occasionally, though, malo doesn't kick in, especially in wines with a low pH, which can be the result of a specific year or grape variety.

The other main bacteria that can be found in wine are acetic acid bacteria. These bacteria ferment ethanol, producing acetic acid and giving off what is known as "volatile acidity" (see *Misconceptions*: *Wine Faults*, pages 78–79), which, if dominant, can spoil the wine and potentially turn it to vinegar.

What conventional producers do differently: Many conventional winemakers actively block malo in order to create a particular style, especially if they are striving for a zippy, zesty white wine. It is about taking a step away from nature toward a stylized product.

Producers block malo by chilling the wine, filtering out or killing the LAB by adding sulfites, or using strains of commercial yeasts like Lalvin EC-1118, which are selected for the unusually high levels of sulfur dioxide they give off during fermentation.

Personally, I believe that blocking malo hampers the development of a wine. It robs the drinker of the full flavor and texture profile that wine is capable of, and those wines that have been purposefully held back in this way often taste as if they have been placed in a straitjacket. Similarly, wines are sometimes inoculated with LAB in order to speed up or control malo.

JUST LET NATURE TAKE ITS COURSE...

Many cellar interventions are related, in some way, to managing the indigenous microflora: to weaken, diminish, or eliminate them, lessen their impact, or compensate for the fact that they aren't working properly. Healthy, active populations of yeast and bacteria go hand in hand with a healthy vineyard. If you start off with great grapes, covered in micro-life, then, as one grower once told me, "Wine just makes itself."

Above and opposite: **All the wines produced by La Ferme des Sept Lunes are fermented spontaneously, and malo is never blocked on any color or style.**

GOING NATURAL

WITH FRANK JOHN

Above: **Frank John owns a 3-hectare, biodynamic vineyard in Pfalz (Palatinate), Germany. He has spent the last 30 years helping others convert to organic viticulture and natural winemaking. Today he works with 164 farms—over 23,000 acres (9,600 hectares)—across Europe.**

You need to think about soils and plants as you would humans. If a human eats all the time, they'll become massively overweight and unable to work effectively. So the first thing you have to do is reduce the weight. And that's what I do with soil. If you're constantly working the soil mechanically and using herbicides and chemical products for fertilization, you're adding oil to the fire. You have to get things back into balance.

Once you start farming properly, the first thing you'll notice is a reduction in the leaf index—the surface of the canopy decreases about 30 percent and the leaves are a lot smaller. The distance between each leaf shortens as well. The whole thing is lighter and the leaves have space to move in the wind. Big leaves end up overlapping, always stuck in the same position. These smaller leaves are actually far more efficient. The plant needs less photosynthesis and less water. It is lighter, healthier, and struggling less. You'll also notice stronger, shorter shoots that don't require as much energy, as well as smaller berries. The fruit might be smaller, but the amount of juice it yields is more or less the same, sometimes even greater. And organic cultivation does not mean lower yields. When you farm properly, plants are healthy and healthy plants are powerful, which means they can carry more fruit, not less.

One of the biggest issues I come across regularly is an overload of phosphorus in the soil. In the early twentieth century, industry grew rapidly, and sectors like iron produced a lot of ash as a by-product that was then used as fertilizer in agriculture, particularly in fruit farming. It was cheap and people were uncertain about the future so they fertilized their soil as much as they could. They treated soil like a bank account—pay in when you have money so that you can take it out when you don't. The result was complete overload, and a huge problem.

I have worked with soils damaged by glyphosates and although it may take five or six years, you can clear them up. I have even worked with soil in Seveso, northern Italy, which was contaminated with dioxins after a factory explosion in the 1970s, and it was hard work but it was possible to get rid of the contaminants. Phosphorous is different—you can't get rid of it, you have to live with it. Vines need very little of it for healthy growth—uptake is only about 2–3kg per hectare. I've worked with soils that have an overload of 1,000 or 2,000kg per hectare—that's 300 years' worth. Imagine you'd been fed 300 years' worth of food in one go. Couple this over-nutrition with strong, fertile rootstocks and you can imagine the result.

And then there's the problem of microbiology for fermentation. Conventional cellars are usually full of lab-bred yeast and bacteria that have been selected to dominate the natural microbiology, and they're everywhere: in the tubes, in the press, on the walls, on the floor.... We usually have to buy in brand new equipment and also start the fermentation in the vineyard itself so that there is no input from the cellar yeasts. This enables us to understand more about the quality and diversity of the yeast in the vineyard, which helps us decide what to do with the green cover and flowering system. Healthy insect populations are essential if you want native yeast for natural fermentation. These yeasts are re-inoculated into the vineyard each spring by the insects, so you need flowering systems that attract lots of different types.

If you're lucky, you might have the yeast you need by the third year but normally it takes five to six years to have enough strength and diversity of yeast that it won't be dominated by the lab-bred imposters in the cellar. Even then it takes diligent hygiene, otherwise the problem can continue for decades. Once the native yeast populations are strong, healthy, and well-fed, they will eventually overpower the old dying system but in the beginning, the strains in the cellar are always stronger.

If you are serious about making the transition to natural wine growing and making, then here are a few tips to get you started:

1 Increase the humus content in the soil. Stop all cultivation (even ploughing) and all additions of synthetic fertilizer. There should be no chemical or physical work done to the soil whatsoever, so it can recover and build up its humus.
2 Use homemade compost. This creates a cycle where nutrients go back into the vineyard they came from, and uses fewer carbon miles.

Above: **Frank runs his own vineyard with the help of his wife and two children.**

3 Whether or not you work biodynamically, spray preparations 500 (horn manure) and 501 (horn silica.) These are great at helping a plant's roots connect with its soil, and its leaves connect with the atmosphere. You will get stronger leaf growth and better fruit that ripens earlier.
4 Use compost tea (best homemade). This is an aerobic water extraction from a good aerobic compost. Apply it to the soil to help increase soil microbiology and prevent anaerobic processes at the same time. It can also be sprayed on the leaves of the vines.
5 Use herbal teas. We make our own from hay and plants that support vine health, like horsetail, or promote soil fertility, like ribwort plantain.
6 Finally, don't use any synthetic chemical sprays on the plants whatsoever. If you spray insecticides, you won't have the right amount of positive insects in your vineyard. If you spray against botrytis, you won't have the right fungi in your vineyard. Anything that destroys soil life will make it extremely hard, if not impossible, to keep things on track. Remember it is a living system.

These six steps need to be done simultaneously, from the very beginning, not step one this year and step two the next, It all has to happen at the same time. You have to remember that you're trying to shock the vineyard. If you want to quit smoking, you don't just swap your pack-a-day for a cigar or even a couple of cigarettes, otherwise you're not free of the addiction. It is the same with the plants. You really do need to take things back to zero before you can see the problem.

THE CELLAR: SULFITES IN WINE

"When I add sulfites, I feel really flat afterwards because I know part of the wine has gone."
(DAMIAN DELECHENEAU, FROM GRANGE TIPHAINE, A NATURAL VINEYARD IN THE LOIRE, FRANCE)

Above: **Gilles Vergé (pictured) and his wife, Catherine, are natural wine growers in Burgundy, France, who make beautiful, sulfite-free wines.**

"We're quite vocal about not adding anything, not even sulfites, to our wines," says Gilles Vergé, a natural wine producer in Burgundy, France. "I guess the fraud and customs officials didn't like that much as, one day, they landed on my doorstep in Burgundy. So began a four-year song and dance that ended in spring 2013. They did every possible test under the sun to try to catch me out. They even used high-resolution Nuclear Magnetic Resonance spectroscopy to work out the composition of my wine. They checked everything—to see if I'd added water, the quality of the grape sugars, the lot. It was an analysis of the sort I'd never seen. Absolutely superb.

"And they found nothing. No traces of anything dubious whatsoever. In fact, in direct contradiction to what many think, they even found 'sulfites: 0.' Yeasts do normally produce a soupçon of sulfites during vinification, but in my wines they found zero. I feel sorry for them, as it must have been a hefty bill."

Gilles Vergé's story is not unusual, as the use of sulfites (or sulphites) is one of the most polarizing topics in the wine world today, not least because of rising health concerns associated with their use in food. Sulfites—or rather a lack of them—is one of the defining characteristics of natural wine.

As Gilles mentioned, yeasts naturally produce small amounts of sulfites during fermentation—usually up to 20mg per liter, although this may be higher with certain strains. However, sulfites can be used at far higher doses by the vast majority of today's winemakers, most of whom argue that not only are sulfites a necessary preservative, but that it is nigh-on impossible to make good, sound wine without them.

Since 1988 (in the United States) and 2005 (in the European Union), all wines containing more than 10mg per liter have to state "contains sulfites" on the label. The real question, however, is *how many* sulfites are present in every liter of wine? For example, a truly natural grower, who does not add any at all, but whose wine naturally contains 15mg per liter, has to mention "contains sulfites" on the label just as an industrial producer would whose sulfite totals might be as high as 350mg per liter. In the EU, sulfite totals are legally permitted up to 150mg, 200mg, and 400mg per liter for red, white, and sweet wine respectively, while a blanket 350mg per liter limit applies in the United States. In short, as things currently stand, we have no idea what we are drinking.

While sulfites can be derived from elemental sources of sulfur, the vast majority of sulfiting agents are by-products of the petrochemical industry. They are manufactured through the burning of fossil fuels and the smelting of mineral ores that contain sulfur. Some of the chemical agents commonly used for winemaking to produce sulfites include sulfur dioxide (SO_2), sodium sulfite, sodium bisulfite, sodium metabisulfite, potassium metabisulfite, potassium hydrogen sulfite, and so on (which are E220, E221, E222, E223, E224, and E228 respectively). As a group they are often generically referred to (by the wine trade) as "sulfites," "SO_2," or (mistakenly) "sulfur."

WHY ARE SULFITES USED?

Sulfites are a common winemaking additive that can take the form of a gas, liquid, powder, or tablet. They may be used at any stage of wine production: as the grapes come into the winery, when the grape juice and wine ferment, or when they are being moved around or bottled. Because of their antimicrobial properties, sulfites are often used at the beginning of fermentation to stun

Below: **The Vergés' vineyard in Burgundy.**

or eliminate wild yeasts and bacteria carried in on the grape berry, so that the winemaker can inoculate his/her chosen strain. Sulfites are also regularly used to sanitize equipment and stabilize wine at bottling. Their antioxidant properties shield wine from contact with oxygen and destroy those enzymes that cause a browning of the grape juice (a similar browning affects apples that are sliced and exposed to the air).

Above: **The purity of sulfite- free wine means it is better for you. As Ghislaine Magnier, founder and ex-owner of Le Casot des Mailloles, who is allergic to sulfites, explains: "The problem with sulfites is that not only are there a lot in many wines, but they are everywhere—preserves, charcuterie, fresh fish—and it's cumulative.**

Opposite: **Transporting hand-harvested grapes to the cellar in small crates ensures the berries remain intact for longer, so reducing the risk of oxidation and the need to use (antioxidant) sulfites.**

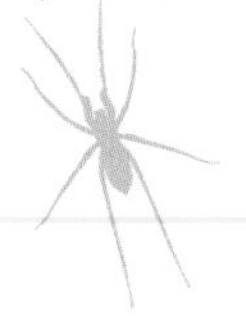

In conventional winemaking, sulfites are often liberally used to control so-called "risk" factors, such as microbes, or to fashion a particular style of wine. Adding sulfites helps to this end. Natural growers, however, welcome diversity and work precisely with the hand that nature deals them each year. They rely on the strength and health of their vineyard to grow great grapes covered with a diverse microflora that will ferment easily and well in the cellar. *Not* adding sulfites helps to this end.

Some natural growers will not add any sulfites at all, while others will add a dash at most, generally at the bottling stage. If they do add any sulfites, it is usually because of commercial realities (perhaps the grower has to release the wines early, for example), trouble with a vintage (because of illness or climatic influences), worries about transport or storage, or nervousness about letting go and believing that the wine will be okay. As Tony Coturri, from Sonoma County, explains, "Wines are a lot hardier than people think. Left alone—without sulfites—they're honestly fine."

To complicate matters further, the use of sulfites is also heavily influenced by culture. "In Germany, Austria, even France, it is a lot more tolerated than in Italy," said the late Stefano Bellotti, of Cascina degli Ulivi, in Piedmont, whose entire commercial production had been sulfite-free since 2006. "In the 1970s/80s, I used to sell 90 percent of my production to organic importers in Switzerland and Germany who would literally force me to add sulfites. My Swiss importer once returned a whole pallet of white wine because, he said, 'it only has 35mg per liter of total sulfites in it and I'm not brave enough to sell a wine like that.'"

"You notice the difference even with the smallest dash of SO_2. Like for like, the wine is duller," says Saša Radikon, a natural grower on the Slovene-Italian border, whose father was one of the first to produce wine without sulfites in his area in recent times. "Between 1999 and 2002 we made two versions of the same wine: one where we added 25mg [sulfites] per liter at bottling and one without. Without fail, the wines with the added SO_2 were 1.5 years behind in terms of aromatic development. Each year we'd show both to professionals and 99 percent of the time, they preferred the no-added SO_2. It is not surprising in a sense since wine needs oxygen in order to evolve at the perfect speed. What's more, we noticed that two years after bottling, the wine with the 25mg per liter added at bottling no longer had any detectable SO_2 present, so it really makes you wonder, why bother?"

A BRIEF HISTORY OF SULFITES

In wine circles, sulfites are often cited as having been in use since antiquity. Yet, when you investigate further, you realize that their use is actually relatively recent. So I thought I should include some of the information on sulfites that I came across while researching this book.

When wine was first "discovered" around 8,000 years ago, somewhere in southern Anatolia (now eastern Turkey) or Transcaucasia (Georgia, Armenia), it didn't happen with added sulfites. In fact, even the Romans, who appeared on the wine scene around 5,000 years later, still didn't use it. "I haven't seen anything definitive," says Dr. Patrick McGovern, Scientific Director of the Biomolecular Archaeology Project at the University of Pennsylvania and also author of *Ancient Wine: The Search for the Origins of Viniculture*. "When we have done tests on residues found in amphorae, we have never found elevated sulfur levels that would suggest they were putting it in intentionally."

Christophe Caillaud, from the Musée Gallo-Romain de Saint-Romain-en-Gal in the Rhône, agrees. "The Ancient World had various uses for natural sulfur. The Romans used it for purification and sanitizing purposes, such as the bleaching of material by the fullers of Pompeii, which Pliny the Elder mentions. Cato mentions its use to combat caterpillars, as well as in a recipe for fixing the coating of wine jars, but it seems that the Ancients missed out on the use of sulfur for the preservation of wine, a practice that only became widespread in the 18th and particularly 19th centuries."

I even enlisted the help of Hans-Peter Schmidt, a natural grower in the Alps, who started his career as an archaeological ecologist, and his conclusions were much the same. "Wine writers always cite Homer, Cato, and Pliny, but none have specific links to wine except for Pliny's *Natural History* (Book XIV, Chapter 25), where he mistakenly cites Cato (*About Agriculture*, Chapter 39). It would take much more time and research to be sure, but I think sulfur was probably not used in either Greece or Rome for wine conservation or vessel sterilization." Instead, the Romans used a variety of other additives (including plant concoctions, pitch, and resin) to correct faults or improve wine of poorer quality. As Columella writes in his *De Re Rustica*, "The most excellent wine is one which has given pleasure by its own natural qualities; nothing must be mixed with it which might obscure its natural taste."

The earliest references I found to "wine plus sulfites" are in German texts from the Middle Ages and relate to barrel sterilization and not to wine preservation. "Sulfur appears to have been introduced in Germany in 1449 and many attempts were made to control it," says Paul Frey, an American organic wine producer who has researched the sulfite question extensively.

This resulted in its being banned entirely in Cologne in the 15th century, for example, on the grounds that "it abused man's nature and afflicted the drinker." At around the same time, in Rothenburg ob der Tauber, the German Emperor decreed against what he saw as "the adulteration of wine and severely restricted the burning of sulfur in barrels. Its use was only permitted for sanitizing a dirty barrel but, even then, you could do it only once, as twice was punishable by law," continues Frey, "and no more than half an ounce of sulfur [per] tonne of wine could be used." This amounted to approximately 10mg per liter, which is a minute dose by today's standards.

What is certain is that by the end of the 18th century, the burning of sulfur wicks—a practice developed by Dutch traders—to shield and stabilize wine in barrel (mainly for transportation purposes) was commonplace. But even then there was hesitation. "I found notes that my great-grandfather Barthélémy kept in around 1868, where he specifically questions the need for using sulfur in wine," explains Jean-Pierre Amoreau, one of Bordeaux's few remaining natural growers, whose estate, Château Le Puy, has been organic for the last 400 years and who has produced sulfite-free cuvées since the 1980s. "But the sulfur he used back then was elemental."

This changed in the late 19th century, as the first oil refineries appeared on the scene and with them the petrochemical industry. Suddenly, sulfites were readily available and this, combined with "advances" in delivery mechanisms in the United Kingdom in the early 20th century—which saw the appearance on the market of liquid and solidified forms known as the Campden Fruit Preserving Solution and Campden Tablet respectively—sealed the future for sulfite use. Now sulfites could be added directly *to* wine, which remains common practice today.

Above: **Many natural growers use no sulfite additions whatsoever. One such is Henri Milan, whose butterfly label red (above) and white wines are totally sulfite-free.**

TASTE: EATING WITH YOUR EYES

"You know what's crazy is that groups of tasters everywhere still consider limpidity as a guarantee of quality or excellence. It's absurd. You only need to pass the wine through a filter and it will be clear!"

(PIERRE OVERNOY, A NATURAL WINE GROWER IN THE JURA, FRANCE)

Above: **While cloudiness in a barrel sample (above) is understood by all in the trade to be part of wine's natural progression, cloudiness at bottling is sometimes (mistakenly) seen as a fault.**

So exclaimed natural wine legend Pierre Overnoy when we met in the Jura in the fall (autumn) of 2013. And, absurd as it may sound, it is the way things are. People *do* eat and drink with their eyes, which is especially problematic for wine. I have often found myself in judging competitions with fellow tasters who want to exclude a wine for being hazy, regardless of its quality. Similarly, growers have for years had problems with local wine boards and export authorities if their wines do not look the way officials expect them to (see *Who*: *The Outsiders*, pages 108–109). As Olivier Cousin, a natural grower from the Loire, in France, says, "It is difficult, because our wines are not filtered, so they have deposits. We have created such a stereotype of the 'perfect' wine that our wines are considered imperfect, and yet we are the ones with perfect wines since they're only grape juice."

Wine is made from fruit, which, when pressed, creates a juice that contains "bits" (pulp, grape skins, live and dead microorganisms, etc.). With time, and the right conditions, these bits settle, and clear wine can be racked off and bottled. Some growers, such as Gilles Vergé in Burgundy, don't bottle for years to ensure their wines are completely settled. Others bottle their wines before the full settling process has finished (often because of cash flow), resulting in wines that can be slightly hazy. Some wines are even purposefully produced with the fine lees kept in the bottle, producing a very cloudy style of wine, as is the case with traditional col fondo prosecco. What's more, over the course of time, even the clearest of proper living wine will develop sediment in the bottle. Most conventional producers fast-track wine settling, using additions and processes (such as fining and filtering) to create the clarity they think customers want. Essentially, there are three options for the wine grower: time, cloudiness, or intervention.

Although cloudiness may occasionally indicate a fault (perhaps a re-fermenting wine, which may produce unpleasant off-flavors anyway), most of the time, it doesn't—just like cloudy apple juice. In fact, some cloudy, natural white wines actually taste *better* if you shake the bottle before opening. The sediment disperses evenly throughout the wine, adding texture, depth of flavor, and overall balance—a bit like adding flesh to a skeleton. Try it. Pour yourself a glass, then gently shake the bottle and taste it again. Have a go with *pétillant naturels*, col fondo proseccos, or even older, unfiltered whites. (Don't do this with old red wines or ports, though, as the sediments are usually larger and are best decanted out.)

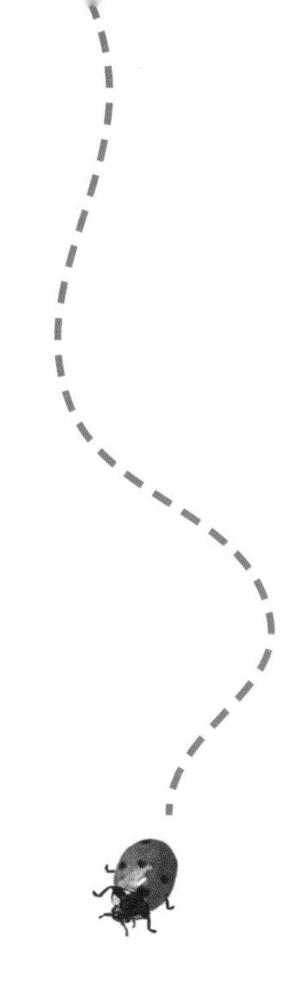

Most of us are formatted tasters. When we hear key words (such as region of origin or grape variety), we taste *inside* a box of knowledge, an important contributing factor of which is the visual. This can be so powerful that it actually alters what we taste in the wine. I once spiked a bottle of riesling with some red, flavorless food coloring and served it blind (i.e. the tasters couldn't see the label) to a group of seasoned, wine-professional friends. Everyone, without fail, thought it was a rosé and even found red berry notes in the wine.

Much of what we taste is so predetermined by what we see that it is really tough to recognize flavors without context. Try this at home: get a friend to cut up as many different nuts and dried fruits as possible into tiny pieces so they're not easily distinguishable. Put a blindfold on and ask your friend to feed them to you one by one. You will see how tough it is to identify them. The visual dominates so much that in order to really connect with flavor we need to be able to step back and focus simply on the mouth. It gets easier with practice, as you start registering individual flavors.

Below: **Try a wine and then decide what you think of it. You'd be surprised how many people have already made up their minds just by looking at the bottle, its weight, its label, or how the drink looks in the glass. The visual is wholly irrelevant to a wine's quality.**

TASTE: WHAT TO EXPECT

"Naturalness is the road, not its end. My goal is to produce profound territorial wines which are only possible without corrections."
(FRANK CORNELISSEN, A NATURAL GROWER ON THE SLOPES OF MOUNT ETNA, SICILY)

Imagine if consistency trumped all else? What would this mean, for example, for unpasteurized brie? Isn't *consistent* Camembert more akin to highly processed cheeses than it is to the oozing cheese that so captured the world's imagination that, when the EU tried banning unpasteurized cheese in the early 1990s, it was met with uproar? As HRH The Prince of Wales put it at the time, "It should strike terror into the hearts of any trueborn Frenchman... and all other people... who find that life is not worth living unless you have a choice of all the gloriously unhygienic things which mankind—especially the French portion of it—has lovingly created."

SO, TRY TO THINK OF WINE IN TERMS OF CHEESE

If we take this as our starting point, then we begin experiencing wine differently, not least because wines with live cultures are very, very different from the sterile, heavily processed varieties mainly on offer. We become far more accepting of the different expressions of its "aliveness." We set different parameters in our minds for what is acceptable. If you've ever tried kombucha—a fermented tea drink containing yeast and live bacteria—you'll know what I mean. The first time you try it, it is surprising. You know it started off as sweet tea, but now it has a distinctly sour flavor profile and is slightly fizzy. But, since you know these traits are meant to be there, you let go and begin to enjoy what you're tasting. This is because the unknown is scary. And, with wine, we have the added complication of thinking that we know lots about it when, in fact, we don't, since most of what we have been drinking is far, far off what we thought it to be. We are full of preconceived ideas.

So, the best way to enjoy natural wines is to try to forget everything you think you know about wine, and start afresh.

Above: **Alain Castex, former owner of Le Casot des Mailloles, who now focuses on his Les Vins du Cabanon, enjoying a glass at *Via del Vi* natural wine hangout in Perpignan, France, where you are sure to bump into some of the Languedoc-Roussillon's best wine producers.**

Opposite: **Orange wines can seem a little unusual at first.**

This page and opposite: **Natural wines are often noted for their elegance and gentleness. The hot climate wines of South-African-cum-Kiwi-turned-Frenchman Tom Lubbe (above right), owner of Matassa in Roussillon, typify these characteristics, as do the cooler climate wines of Burgundian Gilles Vergé (opposite).**

DOES NATURAL WINE TASTE DIFFERENT?

I am often asked whether natural wines taste different. It is difficult to draw general conclusions, given how diverse a group they are, as I am sure you will experience yourself after tasting your way through *The Natural Wine Cellar* (pages 131–205). There is, however, common ground. All fine natural wines, for example, are vibrant (sometimes even a little electric) and full of emotion. They have a broader spectrum of flavors and are usually wines of great purity, often produced without obvious oak additions or too much extraction. They are usually made quite gently, and growers often refer to the fermentation process as an infusion. In fact, as I write this, I cannot help but draw parallels with coffee. Delicious, lightly roasted coffee beans, for example, show far greater aromatic (perfume, acidity) and textural complexity (oils) when percolated rather than being exposed to the quick, harsh extraction of an espresso machine. The result is a drink with a gentleness and elegance that is not dissimilar to natural wine.

Natural wines also tend to have a lovely, salty minerality because of the way in which they are farmed, as the vines are encouraged to cultivate deep roots that engage with the bedrock, processing its minerals through living soils. This proximity to and link with the actual, physical earth means that natural wines also have a far greater array of textures than conventional wines. The liquids have very different tactile sensations, which means that you can almost eat them, and the differences are made all the more stark by the fact that natural wines are neither fined nor filtered but, instead, are given *time* to stabilize and settle.

However, perhaps most important of all is what the French call wine's "*digestibilité*." We (in the wine trade especially) often forget that wine's primary function is to be drunk, making *deliciousness* possibly the most important factor in assessing a wine. And what is certain is that all good natural wines are extremely drinkable. They have a sort of *umami*, or mouth-wateringness, a *je ne sais quoi* that makes you salivate and want to drink more. This isn't perhaps surprising when you realize that most natural growers produce wines that are as much for their own consumption as anything else, and are not made to fit a target market. All in all, natural wines tend to be lighter and more ethereal in nature, and those who enjoy them often remark on their freshness and digestibility.

Because of their aliveness, natural wines behave a little like people. Some days they are more open and generous; others more closed or shy. Some people dismiss this variable living-ness as a lack of consistency, which is a mistake. In good natural wines, the *quality* is always there, but the aromas morph—opening and closing depending on the day or the wine's exposure to air. So, if your bottle isn't tasting as full, for example, as you remember it from last time, leave it until tomorrow, as you might find it suddenly blossoms. Unlike many conventional wines that are identical, day in and day out, year in and year out, until the moment you open them and are then pretty much shot after 24 hours, natural wine is subtly variable and much, much longer-lived once open (see *Misconceptions*: *Wine Stability*, pages 81–83).

NATURAL WINES ALSO TEND TO HAVE A LOVELY, SALTY MINERALITY BECAUSE OF THE WAY IN WHICH THEY ARE FARMED, AS THE VINES ARE ENCOURAGED TO CULTIVATE DEEP ROOTS THAT ENGAGE WITH THE BEDROCK, PROCESSING ITS MINERALS THROUGH LIVING SOILS.

OILS & TINCTURES
WITH DANIELE PICCININ

Daniele Piccinin's 17-acre (7-hectare) vineyard is located in the province of Verona, in Italy. He cultivates various grape varieties, including durella, which is also known as la rabbiosa—meaning "the angry"—on account of its pronounced acidity.

"Most wine growers use Bordeaux Mixture in the vineyard, which is a mixture of copper and powdered sulfur. While this is very effective against mildew, it isn't great for the environment, as copper is a heavy metal that builds up in the soil and groundwater. To eliminate its use in viticulture is tough, however, as your land has to be extraordinarily rich and balanced because fungal infections, like mildew, flourish when there is an imbalance.

We'd been trying to think of an alternative to Bordeaux Mixture for a while and, one day, by chance, I met a man who specialized in the treatment of human fungal infections using plants. We got talking about essential oils and plant distillation, and, combined with my understanding of biodynamic practices, we started creating plant combinations to help restore balance in the vineyard.

That's how our oils and tinctures got started.

EXTRACTION

Plants rich in oil, like rosemary, sage, thyme, garlic, and lavender, can be put through a pot still to extract the oils. Others, though, like nettles, horsetail, and dog rose (*Rosa canina*) are rich in a variety of substances, but aren't oily. *Rosa canina*, for example, is rich in vitamins and is particularly good at helping the body absorb calcium, so is great for menopausal women. However, since you can't extract the essences of these plants in the form of oil, you have to use heat and alcohol to make a tincture instead.

To make a tincture, start by making an *eau de vie*, which means double-distilling some wine in a pot still. The result is a sort of cognac that is about 60–65 percent proof. Macerate the herbs or flowers in this spirit for 60 days, then press and put the liquid to one side.

Dry the leftover solids and start the burning process. We use our outdoor pizza oven where temperatures hit 662–752°F (350–400°C), which reduces the herb solids to a cinder. First, they go black, just like wood on a barbecue, then gray, and eventually bright white. What is really surprising, though, is how salty the white cinders are. I couldn't believe it the first time I tasted them. This is because you've burned off all the water and carbon of the plant, and all that remains are its mineral salts.

Finally, put this ash back into the liquid that you put to one side, and leave to macerate for six months. Your tincture is now ready for use on plants and people.

Burning plants like this, to obtain their essence, is an ancient practice called calcination, which was used extensively in alchemy. In Italy we call it *spagyria*, which means removing all the parts that serve no purpose. By burning away all of the carbon, you're left with the core of the plant—its essence—which is extremely powerful as it is a concentrated version of the plant itself.

Essential oils are potent as well. A drop of pure rosemary essential oil on your tongue, for example, is so strong that you can't taste for the next six hours.

The dosages used of both the tinctures and oils are extremely limited. 30kg of rosemary gives about one liter of distilled water and 100ml of essential oil, which might not seem much, but I only use about five drops, together with 100ml of plant water, in about 100 liters of ordinary tap water each time I treat my vines. You can do about four harvests with a single distillation, so you might distill one set of plants one year and then another set the next.

Our first attempts weren't very successful because, once applied, the concoction didn't stay on the leaf long enough to have a marked impact. But then we added propolis, which is a lot more viscous, and eventually pine resin, too, which is very sticky, so now it's wonderfully resistant to water.

It's a slow process that takes time to perfect, but the parts of the vineyard where we've eliminated all treatments except for oils and tinctures are definitely more resistant than the rest. We do, however, continue to lose some of the harvest each year, plus now we have to fend off wild boar and birds that make a beeline for the oil and tincture grapes in particular."

Above: **Rosehip tincture in *eau de vie*.**

Right: **Daniele Piccinin's pizza oven, which he uses for the calcination of beneficial plants.**

MISCONCEPTIONS: WINE FAULTS

"To make great wine is to flirt with faults."
(PAUL OLD, A WINEMAKER AT LES CLOS PERDUS, IN THE LANGUEDOC, FRANCE)

Above: **Mousiness can result from a wine's exposure to oxygen, and can occur at any time, but particularly during racking or bottling.**

There are some people who claim incorrectly that natural wines are riddled with faults. There are substandard wines, of course, some of which are a direct consequence of mishandled low intervention—after all, natural wine is not immune to bad winemakers. However, truly faulty natural wines are few and far between, and for each one that exists there are many, many more wines that are just perfect.

Below are a few of the most common faults associated with natural wine. Don't be alarmed: none of them are bad for you. The best test of whether the wine is faulty is to decide if you like drinking it. If the answer is "yes," then go ahead.

BRETTANOMYCES Brett is a strain of yeast that can become dominant in the vineyard or cellar, producing a range of flavors that are best described as *farmyardy*. While excess brett will overpower a wine, there is a cultural divide as to whether a touch of it is a positive or negative attribute. In the Old World, there is a much greater acceptance of brett, as it is seen as contributing to part of the style of a wine or adding complexity, but mention brett to an Aussie producer and they will likely freak out. *

MOUSINESS This bacterial infection can develop when the wine has been exposed to oxygen, perhaps after racking or bottling. When the wine reverts back to an anaerobic environment, it settles and the taste disappears. You can't smell mousiness, as it is not volatile at the pH of wine, but the aroma will become apparent when you taste it. Mousiness leaves an aftertaste, reminiscent of off-milk notes, which lingers in the mouth. People seem to be more (myself included) or less sensitive to it. Craig Hawkins, a natural grower in South Africa, explained to me that its presence correlates to a higher pH. **

OXIDATION In a way, this is the most misunderstood of all the faults, since many people unhelpfully use the terms oxidation and oxidative interchangeably. While *oxidation* is a fault, *oxidative* styles are not, and while some natural wines are oxidative in style, not many are oxidized. Oxidative winemaking techniques involve leaving the wine exposed to oxygen, sometimes for years at a time. Natural wines (particularly whites) made with low or no sulfite additions are certainly more exposed to oxygen and so more oxidative. They are generally broader, with hints of fresh nuts and apples, as well as having darker yellow hues. Oxidative notes, though, are not a fault (see *The Natural Wine Cellar: White Wines*, pages 144–61). *

ROPINESS Ropiness is a rare beast and you are unlikely to come across it. Still, in case you do, it occurs when some strains of lactic acid bacteria form a chain, making the wine viscous and almost oily—hence its name in French: *graisse du vin*—while leaving the taste unchanged. As natural growers Pierre Overnoy and Emmanuel Houillon explained to me, all their wines go through ropiness at some stage or another and all eventually return to normal. It does sometimes happen in the bottle, too, but again will resolve itself given time. **

VOLATILE ACIDITY (VA) Expressed in grams per liter, volatile acidity commonly smells of nail varnish. Permitted levels are regulated and cannot, for example, exceed 0.9g per liter for appellation wines in France. But wine is far more than just a set of numbers. It is all about context. A wine can taste perfectly balanced, even with elevated levels of volatile acidity, if, for example, there is a high enough concentration of aromas to back it up. *

OTHER PECULIARITIES If tiny, *perlant* carbon-dioxide bubbles appear when you pour a glass of natural wine, then fear not. Some growers specifically bottle their wines with naturally occurring, residual carbon dioxide because it helps preserve them. Alternatively, carbon dioxide may spontaneously appear if the wine is bottled before all the sugars have fermented, in which case the wine can re-ferment. Again, if the wine tastes nice, don't let it bother you. Alternatively, you can de-gas it by shaking the bottle a little once open.

Tartrate crystals can also sometimes form in the bottle, especially if you chill white or rosé wines for an extended period of time. These crystals are routinely precipitated out in conventional winemaking, but not when natural wine is made. They are harmless and nothing more than naturally occurring cream of tartar. **

So, the next time you hear about faults, ask yourself: what is better, a wine with a touch of brett or volatile acidity or 200 percent new oak? Oxidative notes or clinical, monotone flavor? There is a fine line between complexity and fault. After all, personality means quirks, and I certainly find *personality* far more compelling than bland reproducibility.

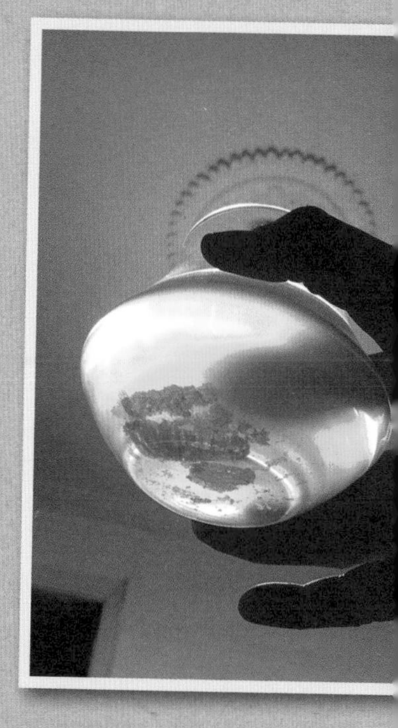

Above: **Harmless tartrate crystals I photographed in a wine I was drinking. I often pick these out to eat them separately. Try it—they are quite lemon zesty.**

KEY TO WINE FAULTS

* A fault that is not unique to natural wine

** A fault that is unique to, or more likely among, growers who work naturally

MISCONCEPTIONS: WINE STABILITY

Above: **Natural wines have the ability to age for decades, thanks perhaps to their internal microbiology.**

Opposite: **Wines stabilize naturally over time, which can sometimes mean they stay in the grower's cellar for years or even decades, depending on the individual wine. In France, this process is known as *élevage*, which rather tellingly is the same word for the action of bringing up children.**

In his book, *Cooked*, Michael Pollan tells the incredible story of Sister Noëlla Marcellino, a cheese-making nun from Connecticut with a PhD in microbiology, who did an experiment to demonstrate that bacteria-rich environments could actually be *more* stable than sterile ones. She made two identical cheeses—one using an old cheese-making barrel, complete with its live culture of lactic bacteria, and another using a sterile, stainless-steel vat. She injected both with E.Coli and found that, while the live cultures resident in the wood quickly colonized the cheese, protecting it from other invaders, the sterile environment proved ripe for unbridled E.Coli reproduction, precisely because it was devoid of defending armies.

It is, perhaps, much the same story with wine. Given time, living wines find their own natural, microbiological balance, making them far hardier than most "protected," preservative-filled, conventional wines. Wine does not need added preservatives for stability. Grapes already possess all the elements needed to ferment and stabilize naturally over time and, if properly produced, this can actually mean wines that, once open, are actually *more stable* than conventional ones, lasting open for weeks in a fridge. Their aromatic profiles change, of course, over the period, but not necessarily for the worse. I have even had bottles that actually taste better one week after opening than they do on the first day.

In any case, natural wines are *living* wines. They are a lot hardier than we think, but, just to be on the safe side, treat them kindly—keep them somewhere cool, away from the stove and direct sunlight, and you'll be fine.

"Living wines are stable even though they might not look like it under microscope. They need to finish their cycle at their own rhythm, so that by the time they go out to clients, they're mature. It's like affinage with cheese: eat it too early and it is far less good."
(NATHALIE DALLEMAGNE, TECHNICAL CONSULTANT FOR VITICULTURE AND WINEMAKING, CAB, LOIRE)

Left: **Natural Loire grower and active environmentalist Olivier Cousin regularly ships his wines by sail using TOWT, whose sailboats are not temperature controlled. Wines are kept cool in the hull of the boat, sometimes spending months at sea.**

NATURAL WINES CAN TRAVEL

Contrary to what some in the wine trade might say, natural wines can and do travel. Wine growers regularly ship wines to far-flung countries, some of which travel in refrigerated containers, while others sit around for months in the heat on docks or on boats at sea.

Stability in wine is fundamentally about time, which means that if you want to cut corners, something has to give—usually either the wines' ageability or its naturalness—and this is only possible with the use of additives or processing. "With our non-young wines, there is no problem whatsoever with stability," says Saša Radikon, an Italian natural grower, who does not add sulfites to his aged cuvées (and whom I interviewed in 2013). "We're releasing the 2007 at the moment, which means it is six years old. It is stable and mature, so that even if it experiences a temperature shock, given a little time, it will recover just fine. Importers sometimes ship these wines in July, for example, in the heat of summer; if they give them two weeks, the wines recover perfectly. But with young wine, it is different. Their structure is more unsettled and less definite, so they can alter for the worse if treated brusquely." Saša overcomes this by adding 25mg per liter of sulfites to his young wines at bottling stage.

AGEING AND NATURAL WINE

Not all natural wines are produced for longevity. In fact, many so-called *vins de soif* are made as early-drinking thirst-quenchers. However, there are plenty of natural wines that are capable of ageing well. I personally cellar natural wines and have had many delicious old examples, including a 15-year-old Casot des Mailloles *Taillelauque*, a 1991 Gramenon *La Mémé*, and a 1990 Foillard *Morgon*. Don't forget, most wines were natural, or at least natural-ish, until very recently (see *Wine Today*, pages 12–15), and many world classics remain so today. I recently drank a 1969 Domaine de la Romanée Conti *Echezeaux*, for example, that was not only still wonderfully fresh and incredibly alive, but also naturally made.

Aged examples are rare, given how minute production is, but you can still get hold of older vintages of Bordelais Château Le Puy, for example, some dating back to the beginning of the 20th century!

Above: **Many natural wines can be happily cellared for years.**

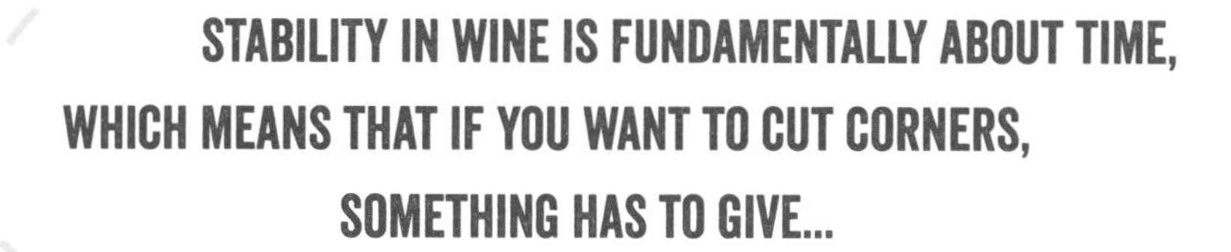

STABILITY IN WINE IS FUNDAMENTALLY ABOUT TIME, WHICH MEANS THAT IF YOU WANT TO CUT CORNERS, SOMETHING HAS TO GIVE...

HEALTH: IS NATURAL WINE BETTER FOR YOU?

"The health of soil, plant, animal, and man is one and indivisible."
(SIR ALBERT HOWARD, A LATE ADVOCATE OF THE ORGANIC MOVEMENT)

Above: **Drinking wine that is full of antioxidants can have beneficial effects on your health.**

Opposite: **A good rule of thumb: red fruits and vegetables, such as black grapes, tomatoes, peppers, and eggplants (aubergines), have naturally high levels of antioxidants.**

Simply put, natural wine contains far less artificial "stuff." For this reason, it's hardly surprising that it might be better for you (not least because many of the additives used in conventional wines are poorly regulated). However, as things currently stand, few studies exploring the effects of wine on health have been undertaken and there have been even fewer on natural wine.

Nevertheless, natural wine aficionados (myself included) often remark on the fact that natural wines give you less of a headache than conventional ones. Personally, I know this to be true because I have not suffered pounding headaches since ditching conventional wine several years ago. But there is a scientific basis to this claim, too. To understand this, let's start with what causes a hangover. Although a hangover is, of course, due to dehydration, what happens in the liver is very interesting. Everything you ingest is broken down by the digestive system and sent to the liver for clearance testing where enzymes process the goods. Healthy components are released into your bloodstream, while toxins are excreted through urine or bile.

Alcohol, or more specifically ethanol, is one such toxin. It is absorbed in the stomach, makes its way to the liver, is identified as a toxin, and is processed for excretion. One set of enzymes in the liver turns the alcohol into acetaldehyde, while another—helped by a compound called glutathione—changes the acetaldehyde into acetate, which is easily excreted by the body. The problem is that, as we drink, the glutathione depletes, leaving more and more unprocessed acetaldehyde to enter the bloodstream. Acetaldehyde is 10–30 times more toxic than alcohol, leading to headaches and nausea as it circulates the body.

In short, glutathione is fundamental to your body's breaking down of alcohol. The problem is that this compound also seems to be highly susceptible to sulfites, as the title to a 1996 paper—"Sulfur dioxide: a potent glutathione-depleting agent"—by the Department of Human Nutrition at the University of Southampton indicates. If correct, this would suggest that natural wine, with

its much, much lower levels of sulfites, is likely to be processed far, far more easily by the liver.

Brand-new research by the University of Rome's Clinical Nutrition and Nutrigenomic Division of Medicine Faculty (which studies how food affects our genes) also supports this idea. As Professor Laura di Renzo, who led the study, explained to me in the fall (autumn) of 2013: "We tested 284 genes before and after the consumption of two red wines—one without sulfites, the other with 80mg per liter in total. We tested the effects of these wines on the genes of the subjects, across different kinds of meal over two weeks. We made two significant discoveries. Firstly, natural wine consumption reduces the amount of acetaldehyde in the blood. This is because of the increased expression of aldehyde-dehydrogenase (ALDH), an enzyme responsible for acetaldehyde metabolism. The other finding was in relation to the oxidation of LDL, the protein that transfers cholesterol around the body, and which is an indicator of oxidative stress of the subject. We found that basically "bad" cholesterol is less when you drink wine without sulfites. These are very, very important results."

What's more, the fruit itself is healthier, too. A University of California Davis study from 2003 found that organic fruit and berries contain up to 58 percent more antioxidant polyphenols, while Dr. Diego Tomasi of the Council for Agricultural Research and Experimentation, in Conegliano, Italy, recently found that grapes that are farmed without synthetic chemicals and without tilling, pruning, and de-leafing, etc., are significantly higher in resveratrol (an antioxidant found in wine) than their conventional counterparts.

Paco Bosco, an enologist, believes that this is because of the vines' adaptability. He spent two years at Dagón Bodegas, a vineyard in Utiel Requena, Spain, for his Masters degree. Dagón does not use any treatments in the vineyard, which has not been pruned or plowed for the last 20 years. The result is grapes with extraordinarily elevated levels of resveratrol. "About double that found in Nebbiolo, which is considered the grape with the most in the world!" exclaims Paco. "Resveratrol is from the family of stilbenes. These are the plant's antibodies, its natural defenses. So, when something attacks it—a fungus or a pest—the plant sends its stilbenes to the afflicted area to fight off the invader." The result is stronger plants, more resilient fruit, and, in the end, healthier wines, especially since processes like fining and filtering, commonly used in winemaking to remove unwanted particles, but eschewed by Dagón (and other natural wine growers), also removes the good stuff like resveratrol as well.

As Tony Coturri, a natural grower in California, told me on a recent visit, "You can't keep putting this stuff in your body. You get allergies. Skin problems. Your immune system collapses. I am old enough now to know people who drank wine their whole life and can't drink it any more, not because of the wine, but because of the additives."

Above and opposite: **Organically grown fruit is naturally healthier, not just because of the lack of pesticides (as is certainly the case with Troy Carter's wild apples, above, which he uses for making cider—see page 129—or the grapes Darek Trowbridge, from California's Old World Winery, collected from a wild vineyard, opposite, while helping Troy to harvest abandoned apples), but, in the case of grapes, also because of higher levels of polyphenols, as experiments at Dagón, in Spain, illustrate.**

WILD SALADS
WITH OLIVIER ANDRIEU

"Every plant has its fungus. Oaks have truffles and vines have their own fungi, too, as do all the other plants in the vineyard. These fungi serve the vines by helping them absorb *oligo-éléments* (trace elements such as boron, copper, and iron) and mineral salts in the soil, which they transmit to the vine. In exchange, the fungus uses the vine to harvest starch since it can't photosynthesize. There's a beneficial exchange. This is what is called a symbiotic relationship.

What is extraordinary is that mushrooms produce filaments in the soil that can link plants to one another, so in the end there is an exchange across the whole plot. A trufficulturist told us just last week about a single mushroom filament that was found, which covered almost an entire forest of several hectares. The trees were all linked. There was a sharing of information via a single mushroom. And we think that it might be the same for the vine.

We try to support this interconnectedness. It means subtle fine-tuning, but we have really noticed a difference. There is a balance that is developing. The vines are more resistant, they're more luminous, and their grapes are wonderful. A little like wild fruit. You can tell they are growing on vines that are not at all stressed.

If you take over a conventional vineyard, there is no symbiosis. No life. You have to start

Clos Fantine, an estate in the Languedoc region in the south of France, is owned by a sibling trio: Olivier, Corine, and Carole Andrieu. They have 72 acres (29 hectares) of vines made up of grape varieties that include mourvèdre, aramon, terret, grenache, cinsault, syrah, and carignan.

by letting other wild plants grow in order to create biodiversity. We have swarms of wasps in our vineyards, for example, and we noticed that when the swarms pass through, we don't have problems with grape moth larvae. Maybe wasps are their natural predator, or maybe they just don't get along. In any case, letting the wild grasses grow has encouraged lots of wasps to patrol the vineyard and we have no troublesome larvae at all.

We have over 30 wild salads and edible plants that grow in among our vines. Some pop their heads up occasionally, some are season-specific, and others are annual. They are all tastiest in spring after the first rains. Here are a few:

Amaranthe (*Amaranthus*): Not indigenous to the area, this used to be cultivated commercially in the 16th and 17th centuries. Nowadays, it grows wild. We eat the first growth of the plant—the very tip of the flower—when it is young and yellow.

Bladder campion (*Silene vulgaris*): Again, the leaves are deliciously sweet, like acacia flowers.

Crow garlic (*Allium vineale*): This looks like normal, cultivated garlic except that it's much smaller and much finer. We use the bulb and cook it in a wine sauce or use the leaves to perfume fish, chopped up like chives.

Common dandelion (*Taraxacum officinale*): All parts of the dandelion are edible, but we particularly like the leaves when they are young and tender.

Marigold (*Calendula officinalis*): The flower is delicious, saffron-like. It adds gorgeous color to salads. You can also use the flowers in soups.

Meadow salsify (*Tragopogon pratensis*): Also known as meadow goat's-beard and Jack-go-to-bed-at-noon, we eat the root of this plant, which is delicious when boiled. Unfortunately, it is becoming increasingly rare.

Navelwort (*Umbilicus rupestris*): Called Venus' belly button in France, this is aptly named since it looks like a navel. Fat and round, the leaves are crunchy and great in salads.

Stone orpine (*Sedum rupestre*): This is a succulent with water-storing leaves and yellow flowers. It tastes a little like shrimp and we eat it *en beignet* (as fritters).

Wall rocket (*Diplotaxis tenuifolia*): We use the flowers as seasoning, in salads, or even for meat. They taste like pepper. Some are yellow, others white. The leaves look like normal rocket.

Wild asparagus (*Ornithogalum pyrenaicum*): This grows around the edges of the vineyards. We eat it mainly in omelets, chopped up into small pieces, or, even better, in a *blanquette de veau* (veal ragout), which is a typical French stew.

Wild leeks or ramps (*Allium tricoccum*): In French we call them *poirots de vignes* or "vineyard leeks." Once blanched, we dip them in vinaigrette.

Wild sorrel (*Rumex acetosa*): We eat the leaves, blanched like spinach."

Stone orpine (opposite) grows in the vineyard (below) at Clos Fantine.

CONCLUSION: CERTIFYING WINE

"It's like asking a high-jumper, who can clear 2m, to jump a bar at 80cm."
(JEAN-PIERRE AMOREAU, THE OWNER OF CHATEAU LE PUY, BORDEAUX, WHEN ASKED ABOUT THE CELLAR REQUIREMENTS IN EU ORGANIC WINE REGULATIONS)

Above: **Two certified natural wine growers among their vines: Didier Barral from Domaine Léon Barral, which is certified organic by Ecocert, but Didier does not mention this on his label or on any other communication...**

August 2012 saw the enactment of new and eagerly awaited European Union legislation governing cellar practices for organic wine, which had not hitherto been included in organic certification at a European level. While being much needed, the actual result represented a setback in many ways. Not only did it make specific allowances for the use of non-organic additives (including added tannins, gum arabic, gelatin, and yeasts), but it also undermined, according to Michel Issaly, former president of the *Vignerons Indépendents de France*, the reputation of organics overall.

Having personally lobbied against the regulation as it was put forward, Michel was horrified by the result. "We knew that the purpose of this legislation was to massify organic wine, so that as many people as possible could produce it, but I didn't see the point of creating an organic label if it ended up being much the same as conventional. When I first saw the file about three or four years ago, I was shocked. How could it allow things diligently protected by organic viticulture to be systematically destroyed by organic vinification? Some non-organic growers I know use far fewer additions, and respect the raw material far more, than their organic-certified counterparts. I worry that, in the end, wine drinkers will start to question what organic really stands for."

This is *the* major shortcoming of certification bodies everywhere—whether organic or, indeed, biodynamic. While regulations may be good at overseeing vineyard management, they all fall short in terms of the winery. What's more, while trying to navigate your way round the dozens of certifying bodies and their particular interpretations is tricky enough, it becomes inordinately more complicated when comparing regulations transnationally—even within the same organization. Take Demeter, for instance, biodynamic's primary international certifying body. In the United States, as in Austria, added yeasts are not permitted, whereas Demeter in Germany allows their use. Similarly, while on the face of it US Department of Agriculture Organic regulations may seem stricter than their European counterparts, since they do not allow 11 additives that are permitted by EU Organic, if you look more closely, the US does allow lysozyme additions, for example, which are prohibited not only by organic regulations in the EU, but also in Brazil and Switzerland, among others.

Consequently, some great growers choose not to be certified. This is partly because, being serious farmers, they go far beyond certification requirements anyway, so cannot be bothered with signing up for additional paperwork duties for an organization whose values they do not share, and partly because of cost. "We looked at some certification, but it was too difficult and too expensive. Some certification bodies required us to give one per cent of our income, and they audit both the winery and vineyard once a year, charging Aus $500/$600 each time. We really can't afford that," says Australian Iwo Jakimowicz of SI Vintners. "I am not anti-certification, but I talked myself out of it. Anyway, why should I have to get certified for *not* putting chemicals on my property when, down the road, they're putting all sorts of stuff on their land and don't have to pay a cent?"

In light of these shortcomings, then, the guarantee provided by a self-regulating grower association such as *VinNatur* offers an impressive alternative. As Angiolino Maule, founder and President of the association, explains, "We are not a punitive association, but an educational one." While actively funding research to help its members to farm better, *VinNatur* is also the only grower association to audit its growers internally by systematically testing samples from its members for pesticide residues.

All in, however, certification is still useful—warts and all. It provides drinkers who may not know a grower's work with a guarantee that they are what they profess to be. Plus, it provides an invaluable structure for the growers themselves. As the biodynamic expert and wine writer Monty Waldin explains: certification closes the back door on exit options, so that when the going gets tough and the temptation to spray rears its head, there's no choice but to carry on.

Above: **... and Yann Durieux from Burgundy's Recrue des Sens, which is certified organic by Ecocert and biodynamic by Terra Dynamis.**

ALL IN, HOWEVER, CERTIFICATION IS STILL USEFUL— WARTS AND ALL. IT PROVIDES DRINKERS WHO MAY NOT KNOW A GROWER'S WORK WITH A GUARANTEE THAT THEY ARE WHAT THEY PROFESS TO BE.

CONCLUSION: A CELEBRATION OF LIFE

Above: **Traditional basket presses are still widely used in natural wine production.**

Opposite: **Harvesters at La Ferme des Sept Lunes in the Rhône; the Lavaysses' donkey at Le Petit Domaine de Gimios in the Languedoc; and Pierre-Jean, Kalyna, and family in their Tuscan vineyard Casa Raia.**

Given that the microbiological life of the vineyard is what enables successful fermentations in the cellar and the creation of wine that is able to survive without a technological crutch, sustaining a healthy habitat in the vineyard for these microbes is fundamental for the natural wine grower. This microbiological life follows the grapes into the cellar, transforms the grape juice, and even makes its way into the final wine in the bottle. Natural wine is therefore, literally, *living* wine from living soil.

In its truest form, natural wine is one that protects the microcosm of life in the bottle in its entirety, keeping it intact so that it remains stable and balanced. However, the production of natural wine is not black and white. As with everything in life, problems arise and commercial realities inevitably inform choices. Natural wine growers can (and do) lose all. Henri Milan, for example, whose celebrated *Sans Soufre* cuvées are drunk around the world, nearly lost his entire year 2000 vintage when bottles and vats started to re-ferment. Therefore, minor interventions—such as a restrained use of SO_2 at the bottling stage—can provide both a sense of security for the grower and a readjustment of the microbial life if aberrations that threaten the quality of the wine begin to occur, while having a minimal impact on the wine.

What's more, while producing wines that are "nothing added, nothing removed" takes enormous skill, awareness, and sensitivity, it isn't always every natural grower's intention. I, for one, added 20mg per liter of sulfites to the first wine I created because I was too scared not to, and, while my wine was definitely not as natural as *Le Blanc* from Le Casot des Mailloles (see *The Natural Wine Cellar: White Wines*, page 151), for example, it was certainly more natural than a standard organic example containing 150mg per liter of added sulfites, as well as industrial yeasts.

"The most excellent wine is one which has given pleasure by its own natural qualities; nothing must be mixed with it which might obscure its natural taste."

(LUCIUS COLUMELLA, 4–40AD, ROMAN WRITER ON FARMING AND AGRICULTURE)

Above: **Many natural growers use indigenous grape varieties (some of which are extraordinarily rare) because they are often the ones best suited to their environment, as well as being part of the natural heritage of their region.**

Natural wine is a continuum, like ripples on a pond. At the epicenter of these ripples are natural growers who produce wines absolutely naturally—with nothing added and nothing removed. As you move away from this center, the additions and manipulations begin, making the wines less and less natural the further out you go. Eventually, the ripples disappear entirely, blending into the waters of the rest of the pond. At this point the term "natural wine" no longer applies. You have moved into the realm of the conventional.

While no legal definitions for natural wine currently exist, various official-ish ones do. These have been set by groups of growers in various countries, including France, Italy, and Spain. These self-regulated charters of quality are far stricter than the regulations imposed by official organic or biodynamic certification bodies (see *Conclusion: Certifying Wines*, pages 90–91). All require the practice of organic farming in the vineyard as an absolute minimum, but also prohibit the use of any additives, processing aids, or heavy manipulation equipment (see *The Cellar: Processing & Additives*, pages 54–55) in the cellar, with the exception of gross filtration, which many tolerate, and sulfites, which varies depending on the association.

The Italian-based *VinNatur*, for example, applies a total maximum sulfite level of 50mg per liter for white, rosé, sparkling and sweet wines, and 30mg per liter for reds. Level 3 of the *Renaissance des Appellations* is also very strict on all aspects of the use of additives and processing, but remains vague on permissible total sulfite levels. The strictest of all, however, seem to be the French natural wine associations, including S.A.I.N.S. (see *Where and When: Grower Associations*, pages 120–21) and the *Association des Vins Naturels* (AVN), neither of which allow any additives whatsoever. For the purpose of *The Natural Wine Cellar* section in this book, all the featured wines comply with *VinNatur*'s totals in order to be able to include a wide range of examples.

Personally, having tasted thousands of examples over the years, I have become less and less tolerant of sulfites. In consequence, most of the wines that I drink are produced without any sulfites, or contain a maximum total of 20–30mg per liter. They are usually neither fined nor filtered.

But perhaps all this is splitting hairs. If you look at the entirety of wine production, and you begin by removing all non-organic vineyards, followed by any others that use added yeast, then those that use enzymes, sterile filtration, and so on, you eventually end up with a very small, precious core of people. Yes, there are differences between a grower who doesn't add anything at all and one who adds 20mg per liter of sulfites at bottling, but, using the ripple analogy once again, while being clearly distinguishable, they are also very close to one another at the center of the ripples.

All in all, true natural wines and their nearly natural friends account for a very small proportion of the wine world. And it is this tiny group that this book celebrates. Not the one-off, lucky cuvée by the likes of me, but the growers who produce exceptional natural wine, year in and year out.

For these growers, what they do goes well beyond the wine itself. Instead, they promote a philosophy, a way of life, which undoubtedly contributes to the profound appeal of their wines to people across the globe. In a disconnected world that worships the Money King, these are people who chose otherwise and who did so well before it became popular. They chose this route out of conviction, a love of the land, and a desire to nurture the most fundamental force of all—life. Be it human, animal, plant, or other life forms, natural growers are primarily, as Jean-François Chêne, a natural producer in the Loire, puts it, about "respecting the living above all else."

Above: **Natural wines are a celebration of life in its purest form. They are farmed at least organically and are made without any additives whatsoever in the cellar. As Camillo Donati, a natural wine producer in Italy's Emilia Romagna, puts it, "For me, it is very simple: natural wine is zero chemicals in the vineyards and zero chemicals in the cellar."**

Overleaf: **Sepp and Maria Muster's living vineyard, in southern Austria, where they produce living wines.**

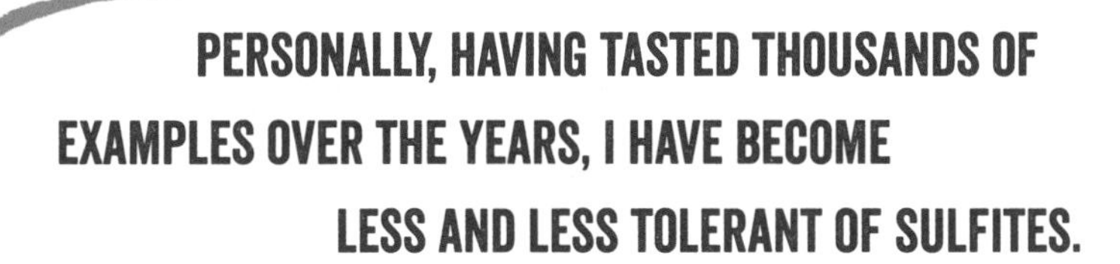

PART 2

WHO, WHERE, WHEN?

WHO: THE ARTISANS

"We don't inherit the earth from our parents,
we borrow it from our children."
(ANTOINE DE SAINT-EXUPÉRY, FRENCH ARISTOCRAT, WRITER, AND POET)

Page 98: **Dandelions in Mythopia's Alpine vineyard. These so-called "weeds" help aerate and fertilize topsoil, thanks to wide-spreading root systems and deep taproots that draw nutrients like calcium to the surface.**

Natural growers come from all walks of life. They may have inherited vineyards from their family or come to farming as a second or even third career. They may be a wild child or a wine nerd, conservatives or kids of the French protests of May 1968. Some fight the system, while others end up as poster boys for it; others choose to stay in the shadows doing what they have always done. But, whether they are radicals or traditionalists, they have all, in some way or other, turned their backs on what the majority of the system sees today as the prerequisites for making wine.

What unites this rainbow army of curious bedfellows above all else is a love of the land. They see themselves as custodians of the natural world and remind us that, done correctly, farming is perhaps the most noble of professions, demanding not only incredible skills of observation, but also a respect and humility that the sublimity of nature necessarily commands.

"We even think about the yeasts and bacteria," explains Jean-François Chêne, from La Coulée d'Ambrosia in the Loire, France. "We try to be close to them, thinking about what they need so that they can be in conditions that are favorable to their work. It's all a state of mind. And it always comes back to the same rule: pick irreproachable raw material and there's no need to ask any more questions."

To grow these "irreproachable" grapes, growers must have an intimate knowledge of their land, which is only possible in the truest sense of being an artisan—a skilled tradesman or woman, working by hand and developing mastery through experience. These growers often work with heritage grape varieties, which are shunned by more commercial growers. "Our goal was to save as many indigenous varieties as possible [from our area] that were on the verge of disappearing," explains natural grower Etienne Courtois, from the Loire, who works with his father, Claude. On the other side of the world, in Chile, natural grower Louis-Antoine Luyt has focused much of his work on the production of pais. This "workhorse" grape was introduced into the country by

Above: **Matassa's Romanissa vineyard at sunset.**

Opposite: **Two Slovenian natural growers—Branko Čotar (right) of Vina Čotar and Walter Mlečnik (left) of Mlečnik—enjoying a glass of wine together.**

Spanish missionaries in the 16th century, but discarded by commercial growers in favor of more fashionable varieties such as chardonnay or merlot, many of which were unsuited to the Chilean climate.

Luyt is also reviving the old Chilean art of fermenting grape juice in a cowhide, with the hairy side facing inward, a practice long since abandoned by all his fellow (adoptive) countrymen. Not only is this, as it turns out, a very efficient mechanism for ensuring healthy ferments, but it is also remarkable for drawing on the kind of ancient wisdom so often dismissed as backward quackery. Be it claypots (such as the Georgian *qvevri/kvevri* or Spanish *tinaja*), orange wines, or even hand-harvesting, natural growers often work using traditional know-how. They are the keepers of heritage practices, which, if discarded, would disappear.

Surprisingly, perhaps, natural growers can also be extremely innovative. They are often already outside the system, so tend to think outside the box, too. Take Californian natural grower Kevin Kelley, for example. Concerned by our reliance on unnecessary packaging, Kevin decided to treat fresh wine like fresh milk,

Soutirage 27 08
Bushido
Bushido
Bushido
La Sorga

setting up the Natural Process Alliance (NPA) project which featured a bottle-exchange program. Every Thursday, Kevin would set off on his "milkround," delivering canisters of wine (straight from the barrel) to customers in and around San Francisco, swapping full canisters for that week's empties—the milk bottles of yesteryear. Sadly the NPA has ceased trading, but the out-of-the-box thinking is characteristic of natural wine producers everywhere.

This originality of thought also extends to ways of living. "In the countryside, we live very independently. We're very self-sufficient," says Olivier Cousin, one of the Loire's natural-wine-grower pin-ups. "Even though, between harvest *et al*, I employ 30 paid people a year, I still do a lot of bartering. We exchange wine for meat, vegetables, all sorts. It's a beautiful community and this solidarity is an integral part of natural wine."

While not all as *bon-vivant* as Cousin, natural growers are often holistic by nature, thanks to their acute appreciation of the subtleties of proper food, health, and life. They may, for example, be as into honey as wine or also cure saucisson or prosciutto at home. The embodiment of this natural philosophy has to be 80-year-old natural wine legend Pierre Overnoy, from the Jura, in France. As well as baking dozens of loaves of delicious sourdough every week for family and friends, he keeps bees, chickens, and an extraordinary collection of grape bunches, preserved in alcohol, that he has picked every year, on July 2nd, since 1990, in order to compare growing patterns. He is as happy to get down and dirty, planting salads or sorting out the plumbing, as he is to talk microbiology or the intricacies of fermentation. He is above all inspiring: warm, gentle, and generous, and his insights sharp and considered.

Unfortunately, there is a mistaken belief that natural growers are laissez-faire or sloppy, which could not be further from the truth. More often than not, good growers tend to be exacting and uncompromising. Antony Tortul, from La Sorga, in southern France, is a case in point. Under a seemingly relaxed air, complete with bushy curls and a wide grin, this young producer runs a very tight ship, where, I suspect, very little is left to chance. He produces some 30 different wines, amounting to 50,000 bottles a year, all of which is done without any additives or artificial temperature control. He is a perfectionist, regularly examining his fermenting juices under a microscope, as well as counting and classifying his yeast populations. He is currently even conducting empirical research in his lab to understand why skin contact facilitates vinification for white wines.

"Our way of doing things couldn't be simpler or more precise at the same time," explains Etienne Courtois. "We make wine like the oldies did—all our presses are 100-plus years old, none are even electric. We cultivate the vines as my father learned to do and his grandfather before him, working as was done in Burgundy over a century ago. Everything is done by hand, which means we walk about 200 to 300km every year just cutting the grass between the rows."

Above and top: **Kevin Kelley and the canisters he used for his NPA project.**

Opposite: **Cuvées by Antony Tortul. After corking, many are sealed with wax, as here, which is a common practice in the natural wine world.**

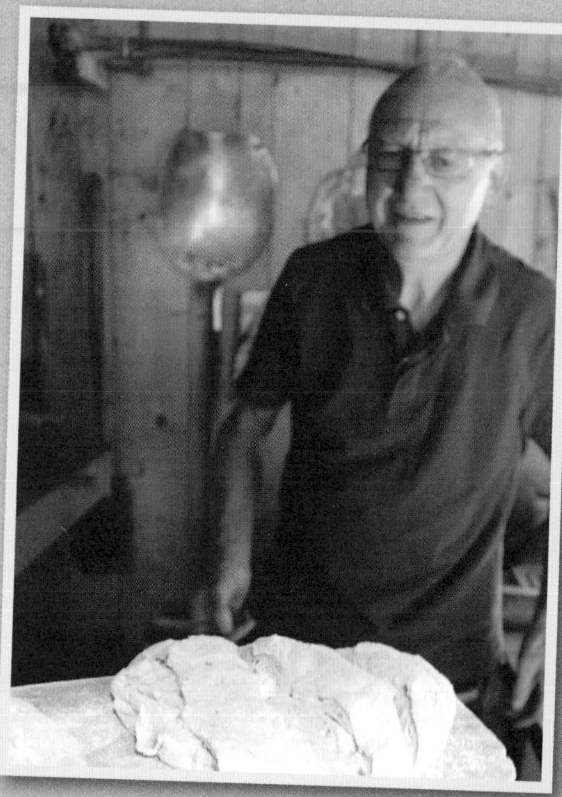

Above: **Natural wine legend Pierre Overnoy, from the Jura, France, baking bread (right), and a bottle of his *Arbois Pupillin* (left), which is today produced by Emmanuel Houillon.**

The results are welcomed with open arms, with entire vintages selling out, and yet, perhaps counterintuitively, the Courtois are actually diminishing their hectarage. "It's about precision," explains Etienne. "My father used to have 15 hectares, whereas today we have about half that amount and we want to reduce it still more. You do often find that successful producers worry about not having enough wine to feed demand. They end up buying other people's grapes to keep up. But it is like having a restaurant. If you start off with a 25-seater that is so popular you have to turn away 50 people every day, it is tempting to say 'I'll just open a 100-seater' and reap the rewards. But a 100-seater is totally different. And it's likely you become just a name on a label."

In many senses, natural wine production obliges a grower to be far *more* exacting than any conventional counterpart. "I am very strict about anything that touches the grapes—pipes, pump, the lot," said the late Stefano Bellotti from Piedmont's Cascina degli Ulivi, who worked sulfite-free. "Three years ago, my press broke down and I was going to have to wait a couple of days for the spare part, so a helpful neighbor suggested I use his. But when I turned up with my 10 tonnes of newly harvested grapes, I couldn't believe my eyes. I always dismantle all my equipment after each press, and steam wash it from top to bottom so that the next day it is spotless for a new pressing. You could lick it, it's so clean. But my neighbor is a conventional producer and adds a lot of sulfites, so he's a lot less fussy about cleanliness. Needless to say, I couldn't take the risk of not adding sulfites that time, as I had no idea what they had been in contact with. So long as everything is clean to start with, then I don't worry about the rest."

This lack of worry is the final piece of the natural-wine-producing puzzle. So much of fermentation is actually unknown that trying to control the process necessarily loses part of its beauty (see *The Vineyard: Understanding Terroir*, pages 40–43). So, growers learn to let go, to trust their instincts. They trust that nature will work its magic because they have done their part to respect it in the first place. It's a real partnership. As Austrian natural grower Eduard Tscheppe from Gut Oggau told me, "It took me six vintages to actually look forward to harvest. On my seventh, I finally knew for the first time that we would be fine. I used to be all nerves, unsure how it would play out, but now I'm in a different place and I love it."

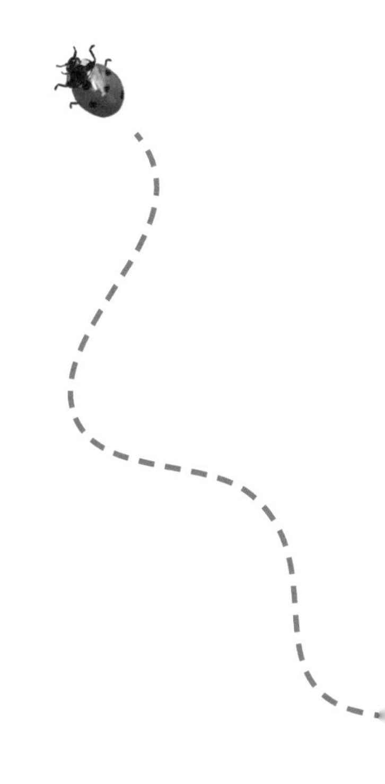

Natural growers don't make wine to a formula or for a market. Instead, what they share is the pursuit of excellence, based on a love of the land and of life, in its most complete and wondrous sense. It is like walking a tightrope without a safety net. As the cellar master of one of the least known natural wine estates, and yet one of its most famous and highly regarded ever, explained to me, "C'est jamais dans la facilité qu'on obtient les grandes choses. It is only when you stand at the edge of the precipice that you have the most beautiful view. It's here, as you risk falling into the void, that you see the extraordinary—overhead, underneath—and it's here that greatness is possible." (Bernard Noblet of the Domaine de la Romanée Conti).

Below: **Austrian producer Gut Oggau's imaginative family of labels.**

HORSES

WITH BERNARD BELLAHSEN

"You know, it's a huge advantage working with animals. First off, there's a real relationship that grows and matures with time.

Then there's the issue that modern agriculture rapes the earth. It forces it to open regardless of whether or not it is appropriate to do so. When you work beside an animal, it's totally different. Animals don't make noise while they work, so you can hear everything that's going on: the plow on the earth, the earth as it opens. You can feel it too since you're there and nothing gets in your way. If it's pouring with rain, your plow will stick and the weight of the sodden earth against the harness will strain your horse. So, you don't plow, which is perfect since drenched soil is vulnerable to erosion and to nutrients being washed away. Similarly, if the ground is too dry, the plow will skate across its surface. Again, you won't plow, which is perfect since cutting topsoil open when it's hot and dry would mean loss of valuable water.

To farm well, farmers ought to consider the state of their soil. They need to work out whether or not it's the right moment to work the land, and animal traction is a great gauge for this. I guess you could do it with a motorized vehicle but you would have to be extremely sensitive to your surroundings. At least when it's an animal leading the way, even an idiot can't get it wrong.

Bernard Bellahsen owns Domaine Fontedicto, a 26-acre (10.5-hectare) farm in the Languedoc, southern France, where he cultivates heritage wheat varieties as well as local terret, grenache, syrah, and carignan grapes. He has farmed organically since 1977, and with horses since 1982.

There's also another advantage. Tractors have combustion engines, whose mini explosions vibrate through the tractor's wheels into the earth. This constant, rhythmic vibration causes the soil to become more compact—like when you squeeze lentils into a jar. You shake and they slot into place. But this pounding eventually pushes air pockets out of the soil, disrupting the soil life beneath the surface. Soon enough the all-important microorganisms that keep your plants healthy and fed disappear. You have effectively squashed your lentils into place. Horses, luckily, don't vibrate. Nor do they explode. I didn't start out working this way but I once watched a farmer ride his horse home at the end of a long day. He lay stretched out on his horse's back. It was such a touching scene, I inevitably wanted the same.

Until the 1950s, Boulonnais Plough horses, from the Pas-de-Calais in northern France, were used extensively in agriculture. They're huge, broad animals, with large muscular chests.

Nowadays, though, most end up at the knackers as sausage meat or pie stuffing. Our Cassiopée, luckily, never made it there. We rescued her at five months old and I worked alongside her for fourteen years. We plowed, harvested, transported, lived side by side, for seven or eight hours a day, weekends included. When you spend that much time together, trust builds and the result is magical. She'd get on with things by herself, no prompting needed. It was extraordinary.

As men, we're not above it all, even though we might think we are. A farmer perched up high in his air-conditioned, automated, glass cabin has a skewed perspective. He's detached. The farmer working with his plow horse, on the other hand, is in the thick of it. He is totally dependent on his companion's strength. He is grounded, literally, since his feet are connected directly with the earth. He sees the state of his land and what needs to be done. He also sees his plants and sees them from a different angle. He is not above but below, or rather inside. He's part of the environment and he feels it.

Working alongside a horse makes you humble. It forces you to listen, to try to work in harmony with your surroundings, and to look at things differently. I can't recommend it enough."

Below: **Bernard Bellahsen and Cassiopée harvesting at Domaine Fontedicto in 2000.**

WHO: THE OUTSIDERS

"I feel huge disappointment. I work so hard to make the purest, cleanest Pouilly Fumé possible, and in the end the industrial wins."
(ALEXANDRE BAIN, NATURAL WINE GROWER IN POUILLY FUMÉ, RECENTLY EXPELLED FROM THE APPELLATION)

"Our wines fail all the time, and it's always the same thing: 'faulty', 'cloudy'... it's a nightmare. We really struggle," explained frustrated South African natural wine grower Craig Hawkins, from Testalonga in the Swartland, when I called him in 2013 to find out how he was getting on with wines blocked at the border. "It's just about box-ticking bureaucracy, but since at the moment there's no natural wine box to tick, they just say,

'You can't send this out.' Once, for example, I stirred my barrels of Cortes just before bottling so it would have fine sediments in suspension. I did it on purpose to have lees to age in the bottle. It was perfectly stable but cloudy. My God, they hated that. My other 2011 wine, *El Bandito*, was sold out before I even bottled it, but we're now in August 2013 and they're still refusing to let the wine leave the country."

Although Craig's wines were sought after by Michelin-starred restaurants in Europe, for years they were regularly rejected by the South African tasters who control export. "Sometimes it gets rejected by three panels, a technical committee, and even the ultimate tasting panel of the Wine & Spirits Board, the PEW committee," continued Craig, "who tell me that allowing my wines for export would harm brand 'South Africa.' In the end, you have a few people deciding what a wine should taste like on behalf of an entire industry. I don't want to lambast or be a rebel, but to make a positive change from the inside so that young creative guys aren't squashed into a hole; too scared to bottle without sterile filtration or fining because of the flack they know they're going to get. After all, it's much easier doing it conventionally. You sterile-filter, finish at 5:30, go home, and have a beer with mates. Nothing more to worry about."

Below: **Stunning, dry-farmed vines in the South African Swartland, where many of South Africa's most avant garde producers live and work.**

Above: **Sébastien Riffault, from Domaine Etienne & Sébastien Riffault, tending his vines in Sancerre, in the Loire, France.**

After years of rejection and hitting up against the system, Craig eventually found a solution, but many others are not so fortunate. In Europe, for example, many growers face eviction from their appellation because of non-compliance with rules on farming or flavor profiles, which have been informed by conventional, standardized, modern practices. Natural grower Sébastien Riffault, in Sancerre, often receives official warnings about the grasses among his vines, while the late Stefano Bellotti, in Piedmont, Italy, who planted peach trees in his vineyard to increase biodiversity, was castigated by officials. According to them, Stefano's actions had "polluted" the land, so it could no longer be considered a *vineyard* and what it produced could no longer be considered *wine*. Absurd as it may sound, Stefano was banned from selling this plot's produce as wine.

Even iconic natural growers sometimes feel the heat. In Burgundy, for example, following a tricky vintage and tiny yields in 2008, Domaine Prieuré-Roch saw its Nuits-Saint-Georges initially refused appellation status because, unlike many of its neighbors, its wines had not been chaptalized and so had significantly lower alcohol levels than the rest. "We have parcels [of land] that have been identified as unique for centuries, if not millennia," says Yannick Champ, joint manager of the Domaine. "Is it really reasonable to think that someone with 20 years' experience can turn that on its head?"

"I've seen grown men break down in floods of tears when they lose their appellation, because it literally means their whole village is turning against them," explains French wine journalist and fellow natural wine advocate Sylvie Augereau. And no wonder, as persecution by the authorities can literally cost you your livelihood. For example, in the fall (autumn) of 2013, biodynamic producer Emmanuel Giboulot, in Burgundy, France, faced prosecution, including a hefty fine and possible time in jail, for not spraying mandatory insecticides on his vines. Other growers have had to close shop following years of persecution.

It is a witch-hunt that has many growers seeking refuge in the "vin de table," "vin de pays," or "vin de France" (or other country equivalents) categories in order to escape the constraints of their now very conventional appellations. Today, however, even these threaten to exclude them, which would make their wines unsellable. "I got in trouble for providing my address on my website," says natural grower Patrick Bouju, from Domaine de la Bohème, France, "because according to the regulations 'vin de table' is not allowed to have *any* geographical mention."

It is difficult to go against the grain. Natural growers take risks on lots of different levels and always put themselves on the line with nature, their conventional peers, and the market. Put bluntly, it takes a pretty brave, remarkable person to stay true to their convictions—this is a big deal in our modern world.

So, next time you pick up a bottle, stop and think for a moment about what it took to get it on that shelf. While they may all look the same from the outside, conventional and natural wines are totally different beasts, and the commitment and effort involved in creating the natural version cannot be overstated.

Above: **Alexandre Bain, Pouilly-Fumé's star grower (whose domaine carries his name), was threatened with losing his AOC status countless times for producing wines that were "atypical." In September 2015, he was evicted from the appellation. He is contesting the decision.**

NATURAL GROWERS TAKE RISKS ON LOTS OF DIFFERENT LEVELS AND ALWAYS PUT THEMSELVES ON THE LINE WITH NATURE, THEIR CONVENTIONAL PEERS, AND THE MARKET.

OBSERVATION
WITH DIDIER BARRAL

"You have to be sensitive to what is going on around you if you want to understand nature. Observation is key.

Everything that happens in the natural world happens for a reason. It has taken millions of years for it to evolve in the way that it has. If things work the way they do, it's because there is a reason for them being that way. It's not accidental or haphazard. It is man who disturbs this balance by creating problems once he starts to intervene. It's really up to him to re-evaluate his practices rather than to question nature's way of doing things. That's why observation is so important—it helps us understand how we can fit into, and work with, what's already in place.

If you drive past a field or vineyard after rainfall, you'll usually notice puddles of water on the ground, whereas if you walk in a forest undisturbed by man, you won't. This is because in vineyards, and in agriculture more generally, we have done away with life. There are no more worms, insects, or other living beings to dig tunnels through the earth and aerate the soil. This is in large part down to the raft of chemical products we use, but even practices like plowing destroy soil equilibrium. It is this equilibrium and the life it supports that makes soil permeable. The key, therefore, is to try to reproduce the equilibrium that exists in forests.

Didier Barral works a 148-acre (60-hectare) polycultural farm in Faugères, in the Languedoc region of France. Half of this area is under vine and includes indigenous varieties like terret blanc and terret gris.

We no longer plow. Instead, we use a Brazilian roller that flattens grasses and weeds between the vines. This protects the soil from the sun, preventing evaporation and helping humidity build up underneath. Without the grasses, the sun would bake the earth, making it virulent and vulnerable. Wind and rain could then blow or wash away valuable clay and humus, and in the end you're left with sand. Keeping grasses intact, even in a sunny, water-competitive environment, is a good idea.

What's more, insects come and live on the grasses, which draws in field mice, shrews, birds, and lots of other animals. All of this ends up back in the soil when it dies, providing a balanced diet for your plants.

In plowed vineyards or, even worse, in those that have been sprayed with herbicide, the vines are entirely dependent on man for support and feeding. When you buy in fertilizers, for example, you are buying a basic mixture of, say, sheep manure and straw. Whereas, if you let wild plants grow in among your vines, a complex web of life

Above: **A cluster of Didier's terret grapes, one of the oldest varieties in the Languedoc.**

Left: **Didier's 50-strong herd of cows includes Jersey cattle (pictured), Salers, and rare Aurochs.**

is created that means a greater complexity of food for your vines.

I used to buy in manure, but, every time I put it on my vines, I noticed that if I lifted the clumps off the soil, there wasn't much underneath. By contrast, if I lifted up dung that my horses had created, there were earthworms, white worms, all sorts of insects wriggling about underneath. The horse droppings attracted life, while the manure didn't. I couldn't understand why. It turned out to be really simple. Deep-litter manure is a mix of urine and poo, and these two excretions do not happen simultaneously in nature. The manure was too strong for worms and insects, so they stayed away. It was at that point that we decided to return grazing to our vineyards, as used to be done in times past. We put our two horses out to pasture in among the vines, as well as our 50 cows. The results were extremely beneficial, as cow pats are warm in winter and cool in summer. This entices earthworms to the surface, regardless of the season, to feed and reproduce. If, by contrast, the earth were naked, cold, or dry, the earthworms would not rise.

What would I say to the young Didier Barral if he started again? Observe. Try to understand what is going on around you, but, most importantly, never go against the grain of nature. You have to be patient and have a keen eye for observation. Spend as much time as possible among your vines, rather than in planes flying the world. Always keep one foot firmly rooted on your land."

WHO: THE ORIGINS OF THE MOVEMENT

"My generation inherited a wave that started over 35 years ago at the hands of a few growers who lived in the line of fire."
(ETIENNE COURTOIS, A NATURAL GROWER FROM THE LOIRE, FRANCE)

Above and opposite: **As the first generation of natural wine icons has begun to step aside, their children have taken on the mantle. Etienne Courtois works closely with his father, Claude, at Les Cailloux du Paradis (above); Matthieu Lapierre is now at the helm of Domaine Marcel Lapierre, working with his mother and sisters (opposite).**

When wine was first made some 8,000 years ago, no packets of yeast, vitamins, enzymes, Mega Purple, or powdered tannins were used. It was made naturally. Nothing was added, nothing removed. Wine *was* natural. However, the need to qualify it as such (i.e. by adding the adjective "natural" before the word "wine") only started happening in the 1980s as a way of differentiating proper wine from the additive cocktail that modern wine had started to become.

Just as the green movement solidified its existence as a result of the Green Revolution, so too the back-to-basics natural wine movement was born out of the intensification of viticulture and an increase in interventionist winemaking. Producers splintered off from the mainstream, questioning the "advances" being adopted by their compatriots, and began experimenting with practices used by their grandparents. Some never stopped being natural, while others became conventional, only to make a U-turn years later.

There is no single individual to whom the movement can be attributed, as examples exist all over the world of people who resisted modernizing trends. They persevered, producing wine in line with their convictions, sometimes wholly unaware of burgeoning natural wine networks in other parts of the world or even on their own doorstep. Life for many was (and in some cases remains) extremely tough, with vineyards regularly vandalized, entire cuvées destroyed, and their methodologies ridiculed by neighbors. "We have it so much easier than my dad did," says Etienne Courtois, a natural grower who works with his father, Claude, one of the Loire's natural wine legends. "It is his generation that did the groundwork… Nowadays there are people who appreciate, listen, and try to understand what these sorts of wine are about. That wasn't the case just 20 years ago when farmers' markets and organic shops didn't exist. Life was much tougher for them."

An extraordinary example of a grower living in inspired isolation is the late Joseph Hacquet, a natural wine visionary who lived with his sisters Anne and Françoise in Beaulieu-sur-Layon in the Loire. Hacquet not only farmed organically,

Above: **La Biancara in the Veneto, Italy, is home to Angiolino Maule, one of the forces behind the Italian natural wine movement. His sons Francesco, Alessandro, and Tommaso work closely alongside him.**

Above right: **Inside the cellar at La Biancara.**

but also avoided the use of any additives in vinification, producing some 50 vintages of no-added-SO_2 wines from 1959 onward. "After the war, natural wine was not only considered anti-conformist but also anti-patriotic," says Pat Desplats, who took over Hacquet's vines with his friend Babass, when the old man was no longer able to tend them himself. "Joseph and his sisters really thought they were alone in the world."

However, fortunately for the spread of the natural-wine contagion, most growers weren't isolated. Some inspired others, causing ripples of interest among neighbors. They then clustered together and sprouted great rhizomes that not only extended regionally, but also nationally, and often, today, internationally. Examples include the Italian-Slovene cluster (started by the likes of Angiolino Maule, Stanko Radikon, and Giampiero Bea) and the Beaujolais cluster (led by the late Marcel Lapierre, along with Jean-Paul Thévenet, Jean Foillard, Guy Breton, and Joseph Chamonard). This French grouping was linked to growers in other regions, such as Pierre Overnoy (in the Jura) or Dard et Ribo and Gramenon (in the Rhône), thanks in large part to the work behind the scenes of two remarkable individuals who proved to be great forces for change: Jules Chauvet (1907–1989) and his disciple, natural-wine-consultant Jacques Néauport. (For more on this key figure, see *The Druid of Burgundy*, overleaf.)

Having begun life as a grower-cum-négociant in Burgundy, Chauvet's fascination with chemistry and the biology of wine soon found him collaborating with research groups throughout Europe, including the Institute of Chemistry in Lyon; the Kaiser Wilhelm Institute (now the Max Planck Institute) in Berlin; and the Institut Pasteur in Paris. Chauvet worked tirelessly, applying scientific method to the problems posed by natural vinification, and his explorations into topics such as the functioning of yeasts, the roles of acidity and temperature

in alcoholic and malolactic fermentations, and the degradation of malic acid during carbonic maceration provided valuable insights for growers opting for the natural route. "I wanted to make wine like my grandfather, but with Chauvet's scientific understanding," explained the late Marcel Lapierre, who was one of the early growers to lead the natural wine charge with their "unusual"' bottles hitting the Parisian natural wine bar scene in the 1980s.

"In 1985, I tasted a Chauvet and shortly afterward one by Lapierre, and that's what triggered it for me," remembers natural Loire grower Jean-Pierre Robinot, from Les Vignes de l'Angevin, who started his wine career as a writer, co-founding *Le Rouge et Le Blanc* magazine in 1983, and then as a wine bar owner with the opening in 1988 of his (bar version of) *L'Ange Vin* in Paris. "We were four or five back then, and I was the last to set up shop," continues Jean-Pierre. "People thought we were out of our minds. We deliberately called it *natural* wine because, while organic, they were also more than that, so we needed a distinction, even though there were actually very few sulfite-free wines available for sale then."

Today things couldn't be more different. The Paris scene has exploded, with dozens of bars, stores, and restaurants stocking natural wines, while New York, London, and Tokyo are not far behind. As American wine writer Alice Feiring told me, "The wines are in huge demand in most progressive restaurants in the major US wine cities—including Austin, New York, Chicago, San Francisco, and LA."

Although a worldwide phenomenon, most natural growers are still located in the Old World hubs of France and Italy. But this is changing, with individual producers popping up in South Africa and Chile and clusters also appearing in countries such as Australia and the United States (particularly California).

Below: **Pierre Overnoy's vineyard in the Jura, France.**

THE DRUID OF BURGUNDY

JACQUES NÉAUPORT

Jules Chauvet (see page 116) is often spoken about in natural wine circles as the father of modern French natural wine. What is much less known is that he did not seek the limelight. Indeed, he spent much of his professional life ferreted away in one lab or another, working alone or with a select group of people in Europe, away from the disdain of the establishment whose feathers he ruffled. It was only after his death that people began to take an avid interest in his work, and this was, in no small part, thanks to one faithful friend who not only brought Chauvet to the forefront of the wine public's consciousness by studying under him, working alongside him, sharing a close friendship with him, and getting his work published posthumously, but also by taking Chauvet's teachings out into the world and putting them into practice in such a way that entire regions blossomed as natural wine hubs. This anonymous shadow is at the root of some of France's most infamous natural wine grower conversions and yet he is probably the least acknowledged, though possibly the single greatest mover-and-shaker in the modern history of natural wine. He is Jacques Néauport, the Druid of Burgundy.

"I was close to Jules until his death in 1989 and I didn't want his life's work eaten by mice, which is what had literally started to happen," explains the 65-year-old vanguard wanderer. "I didn't want that moment of human genius, which was his life, to disappear into oblivion, so I decided to keep his legacy alive. I did my best to get his research published, to write about his life, and to speak of him everywhere I went. Nowadays, the wine world no longer ignores his massive contribution, so I guess I succeeded."

"But, no one knows about your work, all that you have done for wine. Doesn't that frustrate you?" I ask.

"We live in an age of appearances. Everything that 'appears,' exists, and all that doesn't, doesn't. It's just the theatricality of today's society. But, you know, the essential is always done by those you never hear about," Jacques explains.

Here is a man whose tremendous influence contributed fundamentally toward the momentum that gave birth to the movement we know today. He is, in some senses, the Michel Rolland (a wine enologist) of the natural wine world, with a client list that reads like a Who's Who—featuring the famous Beaujolais cluster (including the late Marcel Lapierre, Jean Foillard, Chamonard, Guy Breton, and Yvon Métras), Pierre Overnoy, Pierre Breton, Thierry Puzelat, Gérald Oustric, Gramenon, Château Sainte Anne, and Jean Maupertuis, among many others. Indeed, Jacques even made two sulfite-free white Hermitage cuvées for Chave in 1985 and 1987, which Gérard Chave kept for his private cellar. He worked alongside some of these growers for decades, producing vintages year in and year out (19 with Lapierre, 11 with Foillard, and 17 with Overnoy, for example), and introduced Marcel Lapierre to his friend Jules Chauvet in 1981.

"I prefer not to work with more than 10 growers at a time, as it gets too complex after that," explains Jacques. And yet, there were some years—"easy ones" (his words not mine!)—such as 1996 when he accompanied some 420,000 bottles made without sulfites, a considerable feat then, just as it would be now.

Jacques started out teaching French in the United Kingdom, with most of his salary going on his all-consuming love of wine. Time between teaching assignments was spent on the road, so that by the time he began "doing" wine full-time, some seven years later, he'd notched up 2,000 visits to growers around France. "I always lived for wine, but never wanted to own a vineyard. I wanted to travel, to work with all sorts of different terroirs and varieties."

"I first met Jules back in the spring of 1978. He loved aromas and, after doing a trial with a friend in Pouilly-Fuissé in the early 1950s, discovered that he preferred the complexity of wines with no added sulfites. It was then that he started to vinify without them, but he kept himself to himself because the community had always sidelined him," explains Jacques. "I knew his cousin and since I'd started trying to work with no sulfites in the mid-70s (thanks to partying, hangovers, and witnessing the birth of the UK's Real Ale Movement), it was only a matter of time before I heard about his work. We didn't hit it off too well at first because I arrived at his place late one night, impromptu, acting cocky. I'd been on the '68 barricades, was rebellious, and hadn't yet realized just how exceptional and ahead of everyone else Jules was at the time."

"Natural vinification requires precision. It's like a chain—only as strong as its weakest link. So, you have to be rigorous, and you can't go fast. It takes time. In a way, my role was to reassure growers. There's also no recipe. Three winters I tried, in vain, to write natural winemaking formulas, but it's a hopeless endeavor. The art is in seeing the grapes arrive in the cellar and making the call. The single most important thing is to be organic or, ideally, biodynamic in the vineyard, because you have much, much richer native yeast populations, which I count systematically. Every vintage."

"I've seen unimaginable things," continues Jacques, "among growers who weren't organic or others that, while organic, had fungicide-spraying neighbors that would annihilate their yeasts. But still I managed to make wine. Sometimes it was so tricky; no one else could get the juice to ferment, but I always got it going. Some called it magic, others called it instinct, but, whatever it was, that's how I became known as 'The Druid.'"

"To live happy, live hidden."

(*THE CRICKET*, A FABLE BY CLARIS DE FLORIAN)

WHERE AND WHEN: GROWER ASSOCIATIONS

"I've always been a naturalist, so in 2000 I decided to start surrounding myself with like-minded people."

(ANGIOLINO MAULE, THE FOUNDER OF *VINNATUR*)

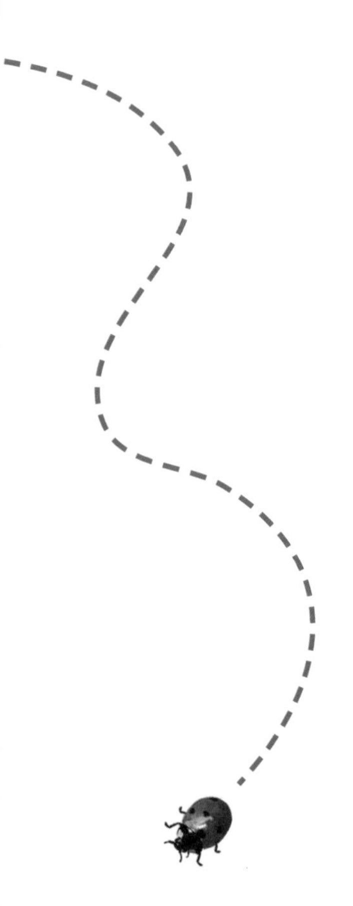

Grower associations are an important part of the natural wine world. In Europe alone, there are more than half a dozen. Most of these are tiny, but there are a handful of larger ones that between them are the movers-and-shakers of the modern natural wine movement. With dozens, if not hundreds, of members in their fold, they provide an extraordinary resource for both growers and drinkers.

Most of the recent advances in wine science have been funded by large enterprise, which means the results tend to focus on issues that are of interest to the conventional, industrial wine community and so are often of little help to growers interested in working naturally. The associations, being grass-root organizations formed by clusters of growers who share a common philosophy, are therefore a great way to exchange ideas, experience, and knowledge. Many of them started out precisely for this purpose.

Associations also provide an opportunity for growers to pool their resources and raise awareness through joint tastings and conferences, where they can share their work with the trade and public at large. The trade, especially importers looking for new wine listings, for example, rely a great deal on these associations and their tastings, but they are also very useful for the end drinker. Given the current lack of regulation, many of these associations have their own charters of quality, which act as a guide to the association's particular philosophy, while also providing drinkers with some basic quality assurance. Here are a few great examples:

S.A.I.N.S., created in 2012, is an association which, although small (only 12 members at present), punches well above its weight, thanks to it being the most natural of all the grower associations. It welcomes only those growers who produce wines without any additives whatsoever in all their cuvées.

VinNatur is a pioneering association because of its innovative collaborations with universities and research institutes, which it undertakes to further our understanding of natural wine in terms of growing and production, as well as

Above: **S.A.I.N.S. is an association of growers who produce totally natural wine—100 percent fermented grape juice, produced without any additions whatsoever.**

its health implications for drinkers. While growers do not have to be certified organic to be part of *VinNatur*, the association does test samples from each of its members for pesticide residues. It works with any growers found with contaminated samples to help them gain confidence and go clean, but, as Angiolino Maule, founder of *VinNatur*, says, "Three strikes and they are out."

The ***Association des Vins Naturels***, in France, is the only grower association, other than S.A.I.N.S., to implement strict limits with regards to sulfite totals. For a grower to become a member, all their wines must be sulfite-free, regardless of style and residual sugars.

With over 200 members, ***La Renaissance des Appellations*** is the largest grower association of note. It was founded by biodynamic farming champion Nicolas Joly. (For more on this natural wine stalwart, see *Seasonality & Birch Water*, pages 44–45.) While not strictly a *natural* wine growers' association (some members' sulfite levels can be quite elevated), many of its members are natural. What's more, it is also the only association where growers have to be certified organic or biodynamic in order to become a member.

Above: **RAW WINE (the artisan wine fair I created in 2012) is the only fair in the world that requires full disclosure from growers regarding any additives or interventions used during winemaking, including sulfite totals.**

WHERE AND WHEN: WINE FAIRS

As awareness of natural wine has ballooned, so have the number of fairs around the world that enable drinkers to meet the heroes behind the bottles. Most of these fairs take place in France and Italy, often organized by grower associations (see pages 120–21) to showcase the wines of their members or by importers in order to show off their portfolios. However, in recent years, lots of independent fairs have also popped up all over the world, from Tokyo to Sydney, from Zagreb to London, each hosting at least one fair where the trade and/or public can come together to meet growers and taste a wide selection of wines under one roof.

La Dive Bouteille, which is only open to wine professionals, celebrated its 15th edition in 2014. It was the first low-intervention wine fair ever created and is, today, the most established in terms of French grower attendance. Founded in the late 1990s by growers Pierre and Catherine Breton, along with 20-odd friends, it was eventually handed over to Sylvie Augereau, a wine journalist and author, under whom it has grown to over 150 growers. "I was militant and took it on as a mission to de-marginalize these guys who were crafting wine as it used to be done, and which they were doing with integrity and a sense of life and community that are rare today. I wanted to defend these ideas and help them be recognized for their work. Speaking to growers at the time, I could see that they were always isolated, so *La Dive* provided a place for them to get together."

Inspired by *La Dive*, I created *The Natural Wine Fair* in 2011, with the help of five importers based in the United Kingdom. The project was short-lived, but paved the way for the creation of RAW WINE the following year, which brings together people (including growers, associations, trade, and public alike) to share their ideas and taste proper wine. With annual tastings in London, New York, Los Angeles, Berlin, and Montréal (and more in the pipeline), RAW WINE is now the largest low-intervention organic, biodynamic, and natural wine fair in the world, and perhaps its most avant-garde thanks to its campaign-like stance

"In the beginning, we were treated like creatures from another planet. Nowadays, we attract buyers from all over the world."

(SYLVIE AUGEREAU, *LA DIVE BOUTEILLE* WINE FAIR)

on full disclosure. RAW WINE aims to move the natural wine debate forward by championing transparency. It is the only fair that asks growers to list any additives or manipulations used in the production of their wines, information that is then made available to the public. It has strict entry requirements and tries to ensure that growers are what they profess to be. Given the lack of clarity to date surrounding the precise definition of natural wine—and given its increasing popularity (plus the temptation for producers to jump on the natural wine bandwagon)—this can be quite a tricky task. However, sticking with fairs that have clearly defined charters of quality, or which have been carefully curated, while not an infallible approach, will likely mean that most growers are compliant.

Other fairs of note, which are more or less natural, include Italy's *Villa Favorita* (organized by *VinNatur*), *Vini Veri*, *Vini di Vignaioli*; France's *Greniers Saint-Jean* (a *Renaissance des Appellations* tasting in the Loire), *Buvons Nature*, *Salon des Vins Anonymes*, *Les 10 Vins Cochons*, *À Caen le Vin*, and *Vini Circus*, to name but a few.

Above: **For RAW WINE 2013, the samples of all the S.A.I.N.S. growers attending the fair were delivered by sailboat into the heart of central London.**

Left: **Brett Redman (pictured here) is the owner/patron of Elliot's in London Bridge. He says, "It's easy for chefs to get their head around natural wine... we care about quality and interesting flavors."**

Right: **With a wide range of styles and price ranges, natural wines can today be found worldwide, from bars and casual eateries such as *Antidote* in London (pictured) to far-flung places like *Soneva Fushi*, an eco-resort in the Maldives that I once worked with.**

Right: ***Noma*, the award-winning two-Michelin-starred restaurant in Copenhagen, won the much-coveted top spot on the *World's 50 Best Restaurants* list for three years running and has served natural wines to diners for years.**

WHERE AND WHEN: TRYING AND BUYING NATURAL WINE

> "For two generations, we have questioned how produce we use in the kitchen is farmed... what we serve in the glass should follow the same philosophy as on the plate."
> (ALAIN WEISSGERBER, CHEF-PATRON OF TAUBENKOBEL, BURGENLAND, AUSTRIA)

"In the beginning we had so much bullshit for it, you can't even begin to imagine," René Redzepi explains. René is the owner-patron of *Noma*, a restaurant that started listing natural wines several years ago. "We were one of the first in Denmark to embrace the idea of natural wine, although just because it is labeled natural, biodynamic, or organic doesn't mean that it is tasty. But the guys who are very good at it, well..." René trails off. "Once you start drinking that sort of wine, it is difficult to go back."

Nowadays, more and more restaurants are choosing to list natural wines because of their precision and purity of flavor. A few years ago, Brett Redman from *Elliot's*—a local, seasonal restaurant in Borough Market, London—went entirely natural with my help. "It's easy for chefs to get their head around natural wine," says Brett, "because we work with produce and we care about quality and interesting flavors. The problem is that most chefs don't understand the winemaking process. Before our list was natural, for example, I thought the winemaker was the most important person in the production process. Now I know it's the farmer." And, like René, Brett believes that once you drink natural wine, you don't go back. "Most of the chefs in the kitchen, within three months of being at the restaurant, only drink natural wine."

Nowadays, natural wines are exported worldwide, so you stand a good chance of finding examples wherever you are. They are such foodie wines that the best choices are often found in restaurants where care can be given to hand-selling them and explaining what makes them so different and special. Some of the world's best restaurants, such as *Fera at Claridges* in London, *Noma* in Copenhagen, *Rouge Tomate* in New York, and *Taubenkobel* in Austria, have wine lists with large natural selections. But great casual dining places abound as well, including *Elliot's*, *40 Maltby St*, *Antidote*, *Duck Soup*, *Brilliant Corners*, *p. franco*, *Naughty Piglets*, *Brawn*, *Terroirs* (and its sister restaurant *Soif*) in London, *Vivant* and *Verre Volé* in Paris, *The Ten Bells* in New York, and *Enoteca Mascareta* in Venice.

Above: **For decades the fine dining, two-Michelin-star Taubenkobel, southeast of Vienna, has championed natural wine, first under the auspices of legendary Austrian food duo Walter and Eveline Eselböck, and today under the guidance of their daughter Barbara and her husband Alain Weissgerber (pictured above).**

Opposite: ***Pas Commes Les Autres*** **in Beziers, southern France, with its 200 or so bins, stocks a good selection of natural wines.**

Above: **A natural wine hangout worth visiting is *The Punchdown* in Oakland.**

Similarly, natural wine bars, which were initially the preserve of Paris, have now spread far and wide. They include *The Punchdown*, *Ordinaire*, and *Terroir* in the San Francisco bay area, *Le Vin Papillon* and *Candide* in Montreal, and *Shonzui*, *Bunon*, and *La Pioche* in Tokyo, among many others. (Japan is one of the largest export markets for natural wines.)

Retail-wise, lots of wine merchants stock the odd example, many inadvertently, and, while you might be lucky and stumble across a bottle in a large grocery store, this is unlikely because the tiny production quantities mean that these stores are generally reluctant to stock them (Whole Foods UK being a notable exception). What's more, as things currently stand, you can't tell a natural from a conventional wine simply by looking at the label, so the best solution, in the UK anyway, is to look online. In France, things are a lot more advanced, with specialty wine shops in most of the major cities, including *La Cave des Papilles* in Paris and *Les Zinzins du Vin* in Besançon. However, New York is close on France's heels, with stores such as *Chambers Street Wines*, *Thirst Wine Merchants*, *Discovery Wines*, *Henry's Wines & Spirits*, *Smith & Vine*, and *Uva* carrying the natural wine flame on the other side of the pond.

PAS COMME lES AUTRES
CAVE A MANGER . VINS VIVANTS

APPLES & GRAPES

WITH TONY COTURRI

Tony Coturri owns an old, unirrigated 5-acre (2-hectare) zinfandel farm in Glen Ellen, Sonoma County, California. He also buys in grapes from nearby organic vineyards and is one of the pioneers of natural wine in the United States. Tony has been farming organically and making wine without additives since the 1960s.

"You might think that places like Sonoma and Napa were always grape-growing regions, but they weren't. The Sebastopol area, west of here, for example, used to be an apple-growing region. But in the early 1960s, it all began to change. The apples being grown were dirt cheap, selling for $25 a tonne. They were worth nothing, financially speaking, so the government came up with a plan and started touting the Gravenstein apple as the future for the area. The Bank of America lent money to farmers to go for Gravenstein and a giant plantation drive ensued.

But the Gravenstein is a soft apple, only good for sauce or juice. You can't store them easily, so they have to be picked and processed quickly, which was a problem since one of the advantages of hard apples was that you could put them in cold storage and deal with them at a later date if you needed to. The whole thing eventually collapsed, ushering in 'the time of the grapes.'

By the end of the 1960s—'67 and '68—the big boom of planting grapes in northern California had begun. By 1972, a tonne of cabernet was worth $1,000, which was huge money at the time. (Nowadays, grapes in Napa can sell for $26,000 a tonne!) So, farmers took out all the apples, all the walnuts, all the pears, it all went.

Apples were out and grapes were in. And the landscape changed.

It was grapes everywhere. Any place you could do it. It went from being a cottage industry to a full-blown manufacturing one, almost overnight. Big tank farms started appearing and big money moved in. Suddenly, the person working the vineyard and the person in the winery didn't own them anymore. Instead, they were working for some person living in New York or LA. Jobs became compartmentalized, and you started getting five winemakers in a single winery, each one working on a different variety. It was a big change. It was the birth of the Sonoma and Napa that we know today, although now it's even more extreme.

It's all monoculture within monoculture. It's not just grapes but one or two varieties only, planted and replanted using identical clones of each. Everybody talks about zinfandel and a bunch of other varieties but, when it comes to planting, cabernet and chardonnay are 90 percent of what everybody puts in. Why bother making merlot when you get more money for cab? Over-ripened, then diluted with water, acidified and 'corrected,' this is the premium Napa product, sold at 100 bucks a bottle.

This total grape domination has had some happy results, though, because there are lots of abandoned apple trees, particularly on the west side of the county. And I'm not talking about a few apples that got missed when farmers were picking—I'm talking tonnes. Troy Carter, from Troy Cider, and I hooked up last year and made cider in my cellar. Ninety percent of what we made was from apples that had fallen on the ground. We just collected them, pressed them, put the juice in barrels, and that was it. The cider made itself, natural yeast and all. Troy took it down to San Francisco and people went crazy for it.

Compared to grapes, though, apples are easy. There isn't the same prejudice surrounding them. There isn't all that stuff that people think wine is supposed to be. People just see it for what it is—cider or fermented apple juice. So, it can be sparkling, it can be cloudy, it can be all the things that wine 'can't' be, and people accept it. There's no 'I'm going to listen to this guy because he knows, and I am going to drink this wine because he tells me to drink it.' There's no *Wine Spectator* magazine reviewing cider."

Right: **Tony's similarly dressed and equally bearded cellar hand (who helps out at vintage), checking the fermenting grapes.**

Below: **Tony's biodiverse, organic vineyard is an anomaly in the context of Sonoma and Napa.**

PART 3

THE NATURAL WINE CELLAR

DISCOVERING NATURAL WINE: AN INTRODUCTION

Above: **Many natural wines are long-lived once open, so why not keep a few bottles on the go in the refrigerator. Have a glass every now and then, and pick your way through the wines as you would a cheese platter. You'll be able to follow the wines' evolution and you might even find that some open up beautifully two days later!**

This chapter is about discovering natural wine for yourself. I have put together a selection of fine and delicious examples to get you started. Think of this list as a mini wine cellar, or perhaps a do-it-yourself starter-kit. It is by no means a definitive list, nor is it a suggestion that these are the best natural wines out there. Instead, I have picked examples because of their diversity and range of flavors, and also because I think that they provide a good cross-section of the types of wine available.

I have mostly stuck to one wine per grower, so that you can get to know as many growers as possible. Each grower does, however, produce other wines as well, so please do explore by all means. You'll find that it pays to be faithful to those you love, as you are not only supporting a grower with extraordinary conviction and commitment to a piece of land, but you'll also start to appreciate the subtleties and greatness of vintage variations.

HOW TO NAVIGATE THE CHOICES

I have separated the wines into six categories: sparkling, white, orange, rosé, red, and off-dry and sweet, with a disproportionately large representation of French and Italian wines, simply because they are the two largest natural-wine-producing countries today in terms of grower numbers. Each category is divided into three color tones. This visual coding is intended to give an indication of the weight or body of the wine. In other words, a "lighter" wine will be paler; a "medium-bodied," middle-of-the-road mouthfeel (i.e. neither light nor full-on) is mid-toned; and a more "full-bodied" (chunkier) wine will be darker. Whites and Reds are further divided by country: France, Italy, the Rest of Europe, and the New World.

I have also included tasting notes on aroma, texture, and flavor in order to help you navigate between the wines, or in case you fancy drinking them with particular dishes and don't fancy opting for potluck. Again, these are not meant to be definitive—natural wines are living products, so they chop and change/open and close/twist and turn a bit like a child's mobile, showing different aromas at different times, and generally being quite moody. Instead, the notes are broad brushstrokes that will give you an idea of the sort of ballpark you're in.

Some of the wines take a while to open up, while others are pretty upfront and less demanding. They are all genuinely thrilling, but some can also be a little confrontational—sort of like experimental jazz.

Finally, there are no scores provided with the wines, as I don't believe in scoring wines, particularly natural ones, given how much they evolve over time. Instead, as Pliny the Elder, who must have foreseen the 100-point system, wisely wrote some 2,000 years ago in *The Natural History*, "Let each person constitute himself his own judge as to which kind it is that occupies the pre-eminence."

THE WINE SELECTION

All the wines listed in *The Natural Wine Cellar* are natural in that, to the best of my knowledge, they have all:

- Been farmed organically and/or biodynamically (or equivalent);
- Been hand-harvested;
- Been fermented with ambient yeasts only;
- Not had their malolactic blocked;
- Not been fined (meaning that all the wines featured are suitable for vegans and vegetarians);
- Not been filtered (with the exception of gross filtration to get rid of flies and such like). Where tighter filtration has been used, I have mentioned it.
- Not had any additives used during winemaking, with the possible exception of sulfites, in which case, sulfite levels do not exceed 50mg per liter for white, sparkling, or sweet/off-dry wines and 30mg per liter for red, rosé, and orange wines, as per the maximums set by *VinNatur* (see page 95), which I used in order to provide as broad a range of wines as possible. The vast majority of the wines listed, though, have had nothing added at all.

A NOTE ON PRODUCTION PRACTICES

All the wines listed come from vineyards that are farmed organically, biodynamically, or a combination of both. While most are certified, some are not (see *Conclusion*: *Certifying Wine*, pages 90–91), but they are farmed, to the best of my knowledge, without weedkillers, pesticides, insecticides, fungicides, or the like. In fact, in many ways, they are all "more than organic," as the growers do more than the bare minimum required by organic or biodynamic regulations.

However, it's important to note that some wine producers don't grow their own grapes, or complement their grape-growing with *négociant* activities, which in some instances can mean buying in batches of grapes that are not farmed organically. As a result, cuvées made from those grapes are not natural, although, confusingly, they may have been made using low-intervention in the cellar. This is worth noting if you decide to explore cuvées beyond those listed in *The Natural Wine Cellar*, as all those I have included have been farmed organically.

Above: **There are no such things as bad or good vintages, just difficult or easy ones for growers. Some years are plentiful, others are not; some sunnier and richer, others wetter and lighter. Since natural wines do not use rectification techniques, vintage variations will be greater and more noticeable.**

TASTING NOTES AND AROMA PROFILES

Tasting notes can be a little reductive, often missing the point about the grower and their philosophy or only capturing a snapshot of a particular wine in a given moment and context, which is especially problematic in the case of natural wines. In short, all the tasting notes in *The Natural Wine Cellar* should be taken with a big pinch of salt.

The aroma profiles, for instance, are intended to give an idea of the flavor profiles to expect and a sense of the wine's freshness, spiciness, etc., rather than a definitive selection of aromas that you will definitely pick out in the wine. There are many reasons for this. Firstly, while certain characteristics of a wine's structure and balance are objective (i.e. acidity, tannins, prominence of fruit versus alcohol, for example), aromas and tastes are incredibly subjective and culturally dependent. If you have never tasted a gooseberry or smelled/tasted hot buttered toast, for example, you will never pick these up in a wine. However, there is probably something else in your life that captures the same feeling or expresses the same taste profile for you (in the case of buttered toast: creamy, malty, with a salty edge). Use this as your point of reference instead. So, if you don't taste what I have tasted, then never mind; if you do, then great.

Our experience of wine is also connected to the emotions of the moment: to the place and friends we are sharing the bottle with, rather than to a specific tasting note or score. So, I suggest you enjoy these wines as you would a great piece of cheese, some pure chocolate, or a delicately perfumed coffee. See how the smells and flavors evolve as you drink, how the wine sits in your palate, how it occupies all sides of your mouth, how different the texture is from one wine to another, and, perhaps most of all, how it makes you *feel*. Is it reassuring or disturbing? Does it bother you or make you feel happy? And so on. Natural wine, in its wholesomeness, is an emotional experience. So, try to approach it with your heart, rather than with your head.

Forget anything you think you know about wine, and just go for it. Take your pick and don't worry too much about what I think.

HOW TO USE THE WINE TASTING NOTES

The sample wine entry on the right is intended to explain the different parts of each of the wine entries in *The Natural Wine Cellar*. Each entry provides key information on the wine's domaine name, country of origin, and grape varieties used, for example.

❶ **Lenticus,** ❷ ***Gentlemant Sumoll***
❸ **Catalonia,** ❹ **Spain**
❺ *Sumoll* ❻ (Rosé)

❼ **Morello cherry | Cranberry | Creamy**

❽ Manel Aviño and his daughter, Nuria, make exciting wines on calcareous soils inside the protected Parque del Garraf, near Barcelona. Focusing on indigenous Mediterranean grapes, his collection of sparkling wine is as diverse as it is prolific. This dark rosé *méthode ancestral* is textured and exuberant, and made from sumoll, a rare local grape that produces terrific wines. Another favorite to look for is Manel's age-worthy still red sumoll, *Perill Noir*, typically released eight or nine years after harvest.

❾ *No added sulfites

❶ DOMAINE NAME

Provides the name of the producer and, as most of the producers in the book also make other cuvées, you can use this name to find their other wines. They're all great producers, so you're in safe hands.

❷ NAME OF WINE

Not all wines have specific cuvée names, which is why the wine name is not always included. (The bottle label has been used as the point of reference.)

❸ WINE REGION

Gives the geographical region of the vineyards and/or cellar. Please note that many natural wines fall into the "table wine" category (or its equivalent). This is often from choice, but sometimes because of problems with their appellations. Consequently, the wine region given here has nothing to do with whether the wine belongs to an AOC, IGP, DOCG, or other designation.

❹ COUNTRY

The majority of the listed wines are from France and Italy, as this is where most traditional, non-high-tech wines are located, although there are also plenty of examples from all over the world. It's likely you'll be able to find a natural wine from a vineyard near you.

❺ GRAPE VARIETIES

The wine lists include wines made from grapes that you will probably recognize, although, as many natural growers work with heritage grape varieties, some of them may be unfamiliar. Don't let this put you off, however, because grapes are far from being the whole story.

❻ COLOR

Gives the color of the wine—whether red, white, orange, or pink—in the Bubbles and Sweets sections.

❼ AROMA PROFILES

Gives some helpful suggestions for the types of aroma and flavor to look out for when tasting the wine. Please note that these descriptions can be highly subjective and will vary from person to person. (For more on this, see "Tasting Notes and Aroma Profiles," opposite.)

❽ SOME BACKGROUND

Provides more detailed information on the wine, whether an anecdote or a particularly striking feature—an appealing texture, for example, or a pearly spritzyness that makes the wine moreish. Also included here, where possible, are further suggestions for similar wines or growers that you may wish to explore.

❾ SULFITE LEVELS

None of the wines in this book have sulfite totals above 50mg per liter (see page 133). In fact, most of the wines have no sulfites added whatsoever. For those where sulfites are added, please bear in mind, though, that these sulfite totals are very, very low indeed compared with most conventional wines.

Note: You may notice that there are no specific vintages in the tasting notes. By conventional standards this is a definite no-no, but I have done this deliberately, to give less weight to specific years. As long as you are dealing with a great maker, who grows or uses great fruit, there is no such thing as a "good" or "bad" vintage (just vintages that are more or less plentiful, or harder or easier for the grower). Or rather, different vintages will be just that, different—although they will clearly be from the same family. Instead, I have tried to isolate a cuvée which is particularly representative of that grower or the region from which it comes. Whichever year you pick will certainly be interesting.

LIGHT-BODIED WINES

MEDIUM-BODIED WINES

FULL-BODIED WINES

Nowadays, the world is awash with brilliant fizz, and it seems on the rise as an increasing number of growers begin to experiment with natural sparkling wines. There are lots of ways of putting bubbles in a bottle, including modern techniques such as the "Bicycle Pump Method" (in which carbon dioxide is injected into a still wine in order to carbonate it) or the "Charmat Method" (where the sparkling is produced in large tanks, rather than bottles)—a method that is widely used today, for example, in the production of prosecco. However, all the wines included here are bottle-fermented, using *traditional* or *ancestral* methods.

BUBBLES

THE TRADITIONAL METHOD

This is, perhaps, the most well-known way of creating fizz—it is used for Champagne, for instance—and is sometimes hailed as producing sparkling wines of the highest quality, which is nonsense since great bubbly can be made using a variety of practices.

Traditional Method sparklings start out as still wines (called the base wine in the trade), which are then bottled, together with yeast and sugar—or, in the case of natural, traditional-method wines, grape juice complete with its native yeasts and natural sugars—to cause a secondary fermentation in the bottle (so creating carbon dioxide). By law, sparklings made using this method then have to be *disgorged*, i.e. the dead yeast cells (known as lees) are expelled.

Although they are the most famous of all the traditional-method sparklings, there are no Champagnes included in my list for the simple reason that making truly natural Champagne is currently illegal. It is a legal requirement to add yeast to start off the second fermentation in the bottle. As absurd as this may sound, a grower cannot, for example, use fresh must (even if it is from the same year, vineyard, or even grapes) for the process. "This practice is legally forbidden in Champagne, despite being legally authorized by the European texts. In Champagne, the '*liqueur de tirage*' [i.e. the mix of yeast and sugar which is then added to the base wine to start the second fermentation] can include saccharose or concentrated, rectified grape must. Not must, *per se*," the Champagne Bureau in London told me in the fall (autumn) of 2013.

"[In Champagne] we need to reach a certain level of pressure inside the bottle which is much higher than for a pet nat, for example, and this is harder to achieve with grape must than through the addition of sugar and yeast," explains Pascal Doquet, President of the *Association des Champagnes Biologiques*. "If we don't achieve the required pressure, or if there are sugars remaining from the fermentation, we are required by law to send the wine for distilling."

It is not just Champagne that finds using must a tricky business. Growers from other regions who are also celebrated for traditional-method sparkling wines experience difficulties using it, too. "I experiment with fresh must but sometimes the fermentation doesn't complete, which causes lower pressure," says Alessandra Divella, a talented young grower-maker in Franciacorta, Italy. "I still use the traditional method with added yeast for the main part of my tirages, but on two barriques I continue to experiment with must each year, hoping in time to become precise enough with it that I can convert completely. I've already corrected a number of mistakes, so I'm positive about the future, but for the moment, given that my production is small, I can't take the risk of producing irregular bubbles."

Since this selection focuses on wines that are made without the addition of yeast and sugar, neither Champagnes nor Divella's sparklings are included here. That is not to say these wines are not worth tracking down. In Champagne, there are terrific growers, including Franck Pascal, David Léclapart, Cyril Bonnet, and Vincent Couche, among many others (the *Association des Champagnes Biologiques* is a great place to go for recommendations), who work very naturally in both the vineyard and cellar.

The most important thing to remember is that growers who work with nature in the vineyard and then carry this philosophy through to the cellar are working naturally as best they can. Certain winemaking styles and regional regulations mean that the producers are not always free to do as they please, but with low additions of sulfites, no fining, and no filtering, they are still capable of creating delicious, complex, live drinks.

THE ANCESTRAL METHOD (AKA PET NATS)

Also known as the *Rural Method*, the *Ancestral Method* is thought to be the oldest recipe for producing bubbles. Fermenting grape juice is bottled straight up so that the carbon dioxide, which is given off as the yeasts convert the sugars into alcohol, is trapped in the bottle. Although beautifully simple, it is, in fact, extremely tricky to get right—bottle late and your sparkling will be flat; bottle too soon and you risk the whole thing exploding. It's a precise art. Growers bottle the juice at a specific density to achieve the right pressure, alcohol, and sweetness. There will probably be slight variations between bottles, and, depending on the stage of their development, some may contain residual sugar. Part of the joy, though, is the wine's evolution in the bottle.

Many of the bottles will have sediments, some more than others, as growers differ in their approaches. Most lightly filter the juice either at bottling or disgorge prior to release.

The resulting wines are nowadays known as Pet Nats—an abbreviation of *Pétillants Naturels*, which, although originally French, is now used ubiquitously worldwide. The best examples are eminently quaffable and are one of the most exciting things to have come out of the natural wine world. They offer extraordinary value in terms of quality, price, and pleasure.

Although pet nats have existed for centuries, they experienced a renaissance with the birth of the natural wine movement at the end of the 20th century. Today they have taken the world by storm, moving from novelty drink to being an extremely successful category of wine in their own right.

Many natural wine growers today make a pet nat of some description, often in small quantities of around 3,000 to 4000 bottles. Many are early-drinking, juicy bubbles but, as is so often the case with wine generally, they invariably benefit from a little bottle age. They are available in all the different colors—white, orange, pink, and red—and versions of each are included in this selection.

Opposite: **Costadilà's col fondo sparkling wines in the Veneto, Italy, are part of a new wave of producers creating proper bottle-fermented, prosecco-style wines. While its founder, Ernesto Cattel—who spearheaded the col fondo renaissance in Prosecco—sadly passed away, his team continues his work today.**

Right: **Pétillants naturels (or pet nats)—natural sparklings—have exploded over recent years, and with good reason: they are some of the most exciting, easy-drinking wines around.**

LIGHT-BODIED BUBBLES

Les Tètes, *Tète Nat, Vin de France*

France

Loin de l'oeil, semillon, mauzac, grenache (White)

Granny Smith (apple) | Honeydew melon | White flower blossom

Natural wine is today a bit of a buzzword—with bottles selling out like hot cakes—and as its popularity has grown, so too have the number of *négociants* (producers that buy in grapes or wine and then bottle them under their own label). Even growers with their own farms, who may have started out solely bottling their own grapes, have expanded their ranges to include *négociant* bottles as a way of increasing production and revenue—an excellent source of extra income, especially following a bad vintage, for example. This is, of course, a positive development since it means the demand for natural wine is on the increase, but it is one that needs more control. It is often difficult to trace the origin of the bought-in grapes, making it all too easy for producers to use cheap, conventionally farmed grapes from the local co-op, which they then flog as "natural wine," especially if they are trusted natural wine growers themselves.

Enter Les Tètes, a *négociant* project created by four friends to make and sell pétillants naturels, although they have now extended their range to include still wines, too. Philippe Mesnier, one of the partners, regularly travels to the vineyards from which they buy fruit, working closely with the grape-grower and controlling the quality of the farming, including deciding harvest dates, for example. They test all their wines for pesticide residues (including glyphosate), an expensive test but one that they deem necessary even if the vineyards they work with are certified organic. "We have a commitment to the drinker to deliver honest wines, so we need to make doubly sure that the grapes we buy are farmed cleanly," explains Philippe. "Too many *négociants* exist in a gray area and we don't want to do that."

Their *Tète Nat* pet nat uses grapes from various parts of France (Gaillac, Bordeaux, and the Rhône) and is bottled once the residual sugar in the fermenting must is low (around 15g). This is then fermented quickly to create, as Philippe explains, a fine bubble that needs little *élevage* to achieve stability and can be disgorged as early as February the following year. The result is a racy fizz that makes for a deliciously refreshing apéritif.

*No added sulfites

Quarticello, *Despina Malvasia*

Emilia-Romagna, Italy

Malvasia (White)

Honeysuckle | Lychee | Conference pear

Like Cinque Campi described on page 143, Roberto Maestri, who owns Quarticello, is part of a proper lambrusco renaissance sweeping the Emilia Romagna region at the moment. Lightly fizzy, floral, and with hints of apricot, it shows great intensity of crystalline aromas. Extremely precise and linear.

*Low levels of sulfites added

La Garagista, *Ci Confonde*

Vermont, USA

Brianna (White)

Pollen | Fresh date | Peach

Husband-and-wife team—Deirdre Heekin and Caleb Barber—began life as dancers, and are today biodynamic farmers-cum-restaurateurs-author-baker-winemakers! Their work turns conventional wine "wisdom" on its head, as they work primarily with hybrid varieties: la crescent, marquette, frontenac gris, frontenac blanc, frontenac, brianna, and St. Croix, which are basically crosses between various grape species, including *vinifera* (traditional European wine grape varieties) and some hardier wild, native American species (such as *riparia* and *lambrusca*). Originally bred to suit the climate, hybrid varieties have fallen so far out of conventional wine favor that most wine professionals have never even tasted them. Hybrids have very unusual flavors and textures, and tasting through Deirdre and Caleb's range will certainly take you out of your comfort zone because they are so unusual, so wonderfully refreshing, and so darn good.

*No added sulfites

La Grange Tiphaine, *Nouveau Nez*
Montlouis, Loire, France
Chenin blanc (White)
Quince | Water apple | Mirabelle

This 25-acre (10-hectare) estate, created by Alfonse Delecheneau at the end of the 1800s, is today in the hands of his great grandson, Damien, and his wife, Coralie. They craft a great range of wines using sauvignon blanc, cabernet franc, and, of course, chenin blanc, the star grape of the Montlouis region. One of my favorites is their dangerously drinkable pétillant naturel. Precise and understated, it is full of pleasure—an elegant wine.

*Low levels of sulfites added

MEDIUM-BODIED BUBBLES

Frank John, *Riesling Sekt Brut 41*
Palatinate, Germany
Riesling (White)
Sourdough | Acacia blossom | Lanolin

The jovial Frank manages his 8-acre (3-hectare) vineyard (and 400-year-old Renaissance cellar) with the help of his wife and two kids. He also consults for hundreds of vineyards across Europe, helping them farm organically and make wine naturally (see pages 62-63). He's a stickler for detail, deep cleaning and limewashing his cellar every year before harvest to ensure the yeasts that start the fermentation come from the vineyard and not the cellar—true vintage expression. His traditional-method *Riesling Sekt Brut 41* (which is made by adding fresh must to his still wine to restart the fermentation) goes through a long ageing process of 41 months (hence the name), and the result is a deliciously smoky, savory, complex bubbly.

*Low levels of sulfites added

Lentiscus, *Gentlemant Sumoll*
Catalonia, Spain
Sumoll (Rosé)
Morello cherry | Cranberry | Creamy

Manel Aviño and his daughter, Nuria, make exciting wines on calcareous soils inside the protected Parque del Garraf, near Barcelona. Focusing on indigenous Mediterranean grapes, his collection of sparkling wine is as diverse as it is prolific. This dark rosé *méthode ancestral* is textured and exuberant, and made from sumoll, a rare local grape that produces terrific wines. Another favorite to look for is Manel's age-worthy still red sumoll, *Perill Noir*, typically released eight or nine years after harvest.

*No added sulfites

Gotsa, *Pet' Nat'*
Georgia
Tavkveri (Rosé)
Wild strawberry | Rhubarb | Cacao bean

Former architect Beka Gotsadze is a larger-than-life character, with a warm, booming presence. After searching for years, he decided to plant vines in what used to be one of Georgia's ancient wine-growing areas, off the Tbilisi-Armenia road, south of the capital. He's the only grape grower for miles around, with locals having swapped vines for sheep when the Soviets took over. Beka chose eastern Georgia for grape-growing, as its fertile soils meant higher yields. Beka ferments and ages his wines in *qvevri* (or *kvevri*)—a wine "technology" included on UNESCO's List of the Intangible Cultural Heritage of Humanity—and uses gravity to move his wines around his hilltop cellar. A dash tannic, this vibrant, deep pink, flavorsome pet nat, made without skin contact (unusual for Georgia), is awesome—a huge achievement considering that this is Beka's first go at making natural bubbles. It is a testament to how resourceful and exacting he is, given the uncompromising attention to detail needed to make a success of it. Well worth seeking out.

*No added sulfites

Costadilà, *280 slm*
Veneto, Italy
Glera, verdiso, bianchetta trevigiana (Orange)

Crushed rice | Peach | Ginger

Orange and fizzy, this floral wine is creamy with tannins, as a result of the 20- to 25-day maceration on the skins, which takes place without any temperature control. Second fermentation is carried out in the bottle by adding fresh must (complete with its wild yeasts) from dried, pressed grapes of the same harvest. There are no additions whatsoever.

*No added sulfites

Domaine Breton, *Vouvray Pétillant Naturel Moustillant*, Loire, France
Chenin blanc (White)

Propolis | Cinnamon | Baked apple

Catherine and Pierre Breton (creators of *La Dive Bouteille*) make a cracking range of fizzes (and stills). This is my favorite of their range, with its delicious baked-apple and cinnamon edge, and super-creamy mousse.

*No added sulfites

Vins d'Alsace Rietsch, *Crémant Extra Brut*
Alsace, France
Pinot auxerrois, pinot blanc, pinot gris, chardonnay (White)

Gingerbread | Ripe persimmon | Vanilla bean

Like many Alsatian producers, Jean Pierre Rietsch makes an extensive range of styles. A shy, playful man, his wines are delicious—some with a touch of sulfites, others without. I particularly love this *Crémant d'Alsace* (the term used to describe traditional-method sparklings from this region). Creamy and opulent, it was made without any dosage or sulfite additions, and the second fermentation in the bottle was started with must from his 2014 vintage.

Another couple of must-tries are his oranges made from gewurztraminer and pinot gris—both are made with skin contact, which adds a savoriness and tightness to grape varieties that can sometimes be a little over-the-top.

*No added sulfites

Les Vignes de Babass, *La Nuée Bulleuse*
Loire, France
Chenin blanc (White)

Mimosa | Honey | Ripe Williams pear

Sébastien Dervieux (aka Babass) created his own domaine after working at Les Griottes with Pat Desplats. It is Sébastien who today cares for Joseph Hacquet's old vineyard (see *Who*: *The Origins of the Movement*, page 116). A dark yellow, sparkling wine, with a dash of residual sugar, it has a honeyed nose and creamy texture. A few darker spices, too, as well as great concentration—a trait that many well-managed vineyards often have.

*No added sulfites

FULL-BODIED BUBBLES

Movia, *Puro*
Brda, Slovenia
Ribolla gialla (White)

Peach blossom | Linseed | Macadamia nuts

Movia has been around since 1700. Aleks Kristančič, an energetic man, has championed bubbly *sur lie* (on the lees) for many years, saying that they are essential for keeping wine alive. He makes delicious, long-lived sparkling wines. Base wines are aged for a number of years before being bottled with fresh must. These are released undisgorged, and Aleks advises people to disgorge them themselves before drinking. Personally, though, I prefer not to, as I enjoy the added texture of the fine lees when they are mixed in with the rest of the wine—just be careful not to shake the bottle too vigorously before opening.

*No added sulfites

Casa Caterina, *Cuvée 60, Brut Nature*

Franciacorta, Italy

Chardonnay (White)

Golden Delicious (apple) | Brioche | Sesame seed

Owned by the Del Bono family, who have lived and farmed in this area of Italy since the 12th century, this 17-acre (7-hectare) estate grows dozens of different grape varieties and produces micro-cuvées, around 1,000 bottles each. Matured for almost five years (or rather 60 months—hence its name) on its lees, the *Cuvée 60* has developed great complexity of autolytic bready notes, while retaining a beautiful, almost lemon-balmy, freshness. It has a great creamy texture. Ripe. Round and opulent, with a sensation of sweetness.

*No added sulfites

Les Vignes de l'Angevin, *Fêtembulles*

Loire, France

Chenin blanc (White)

Bready | Medlar | Greengage

Jean-Pierre Robinot, one of France's earliest natural wine supporters, started off as a wine writer. After co-founding *Le Rouge et Le Blanc*, a wine magazine in France, he opened one of the first natural wine bars in Paris, in the 1980s, before going country and deciding to grow grapes himself. Deep, complex, and bone-dry with yeasty brioche notes and an almost metallic minerality. It is verbena-esque, although very dry.

*No added sulfites

Camillo Donati, *Malvasia Secco*

Emilia-Romagna, Italy

Malvasia (Orange)

Damascus rose | Lychee | Marjoram

Camillo's wines are bold and exciting, and this fizz is no exception. It is full-on. Forty-eight hours on the skins helps to give it a "bitey" texture and highlights the floral aromas of the malvasia. When I tasted it, it showed beautifully even two days after opening. I drank it with some super-simple spaghetti—olive oil, sage, and old, crunchy Parmesan cheese—and it was delicious.

*No added sulfites

Capriades, *Pepin La Bulle*

Touraine, Loire, France

Chardonnay, chenin blanc, menu pineau, petit meslier (White)

Ripe melon | Brioche | Carambola

Pascal Potaire and Moses Gaddouche are, if you like, the "mac daddies" of the pet-nat method. The *méthode ancestral* is all they do—and they do it to perfection. Ask most French *pet-nat* growers who their heroes are and, more often than not, they'll cite Pascal and Moses. This cuvée, released after three years of *élevage*, is one of the more serious of their range. Opulent and ripe, it has incredible weight and concentration. They also produce others, such as their *Piège à Filles*, which are much lighter in style and great apéritifs. Utterly delectable wines.

*No added sulfites

Cinque Campi, *Rosso dell'Emilia IGP*

Emilia-Romagna, Italy

Lambrusco grasparossa, malbo gentile, marzemino (Red)

Cassis | Black olives | Violets

Although red sparkling wines are (unfortunately) pretty rare, this area of Italy has produced some gorgeous examples, including this one. Tannic and full-bodied, with a refreshing, crunchy acidity, it shows dark fruits typical of serious lambrusco. And it is so savory, it is almost meaty. Works with fatty foods really well. Only 3,000 bottles were produced. Cinque Campi's entire range is sulfite-free.

*No added sulfites

LIGHT-BODIED WINES

MEDIUM-BODIED WINES

FULL-BODIED WINES

If you drink conventional whites, then this is the category that is most likely to surprise you, as natural whites tend to be fuller in style and more individualistic (or unusual) than their conventional counterparts. They offer a greater variety of flavor profiles, but also much less of the zippy tartness associated with some types of conventional white wine.

WHITE WINES

MAKING WHITE WINES

White wine is generally made using a technique known as "*pressurage direct*" (that is, by pressing grapes and fermenting the run-off juice without skins or pips, or allowing a few hours of contact at most). This means that white wines don't benefit from the naturally occurring tannins and antioxidants (such as stilbenes) that are extracted from grape bits (the skins, pips, and stems) during extended macerations and which help to protect the juice. As a result, white wines tend to be a lot more fragile during the winemaking process than reds or oranges, and need more care.

Since natural growers do not resort to the usual artillery of sulfites and lysosomes *et al* that conventional producers use throughout the winemaking process, they can't succumb to fears of exposing their must and wine to oxygen. In some ways, I think it's at this initial stage of the fermentation process that growers have to show the most faith in what nature can achieve. They have to trust that, as long as they have harvested healthy, microflora-rich grapes, they have nothing to fear and that what looks, for example, like a browning of the juice will over time revert to the paleness of a white, or that the yeasts and bacteria will do their job to the end and that the wine will clarify naturally over time.

WHY DO NATURAL WHITES SOMETIMES TASTE SO DIFFERENT?

Openness to oxygen does, however, inform the wine's taste and texture, and it is this difference that causes the greatest commotion among conventional wine drinkers. Indeed, most of the criticism generally leveled at natural wine, and the way it tastes, comes from white wine. You might, for example, hear people say that it tastes like cider or that the wine is oxidized, as people sometimes (mistakenly) describe oxidative notes in this way. Some natural white wines

Left: **Julien Peyras from the Languedoc in France is a very promising young producer. His rosé is featured on page 175.**

are indeed oxidized and do taste of cider, but you would be surprised how often tasters use this as a blanket term to describe a panoply of flavors.

It is true that when you drink a natural white, particularly a wine made without any sulfites at all, the combination of texture, ripeness, and amplitude in the palate sharply contrasts with conventional counterparts made in a temperature-controlled environment, with added yeasts, sterile-filtration, etc. Take sauvignon blanc, for example. Truly international and very fashionable, it is widely enjoyed for its vibrant, overt citrus and gooseberry notes, and zippy acidity. These are what most people consider to be the defining characteristics of wines from this variety of grape. Now, imagine that, actually, there is another side to sauvignon blanc: a darker, more serious side to this seemingly fluffy, shallow grape. A sauvignon blanc that, if left to mature fully and made from balanced yields and organic vines, would show luscious acacia honey notes and a round and creamy mouthfeel. The unexpected nature of this full broadness would come as a shock and all you'd likely taste would be the fact that, in comparison with the zesty, tart, watery white sauvignon you normally drink, this *seems* oxidized. To better understand what I mean, consider the difference between an unripe, hydroponically grown winter tomato versus one you bought at the local market on a summer vacation in Sicily. Now, imagine you'd spent your life eating hydroponic tomatoes and, suddenly, you bite into a properly grown and ripened one. You wouldn't know what had hit you. The intensity of flavors would be overwhelming and, in comparison to the bland, tarter version from the greenhouses of Rotterdam, it could also taste more "oxidized" or a bit more sun-dried-tomato-esque. It is simply that the range of flavors sits in a completely different part of the flavor spectrum. This is not to say that there are no oxidized natural wines, far from it, but there are a *lot* fewer than people make out.

Another layer of complexity is introduced by the impact of malolactic fermentation (also known as "mlf" or "malo"—see *The Cellar: Fermentation*, pages 57–61). When you don't work with sulfites during fermentation, wines (of all colors) usually go through malo. This secondary fermentation usually takes place after the alcoholic fermentation and sees bacteria (good bugs, not bad) convert malic acid, which is naturally contained in the juice, into lactic acid. This transformation of acids fundamentally changes the texture and flavor of the wine, since lactic acid is a softer, broader type of acidity than malic. What's more, since the bacteria responsible for malo are naturally contained in the environment, their presence in a particular year is wholly dependent on the conditions of the vintage. As a result, as Jean-Pierre Amoreau from Château Le Puy explained to me in September 2013, "You cannot talk about *terroir* if you block malo."

More often than not, the conventional camp sees malo in whites as an undesirable trait that must be curtailed, actively blocking it to create wine of a particular style, such as one with a zippy mouthfeel. This is done by destroying the bacteria responsible for malo by chilling the wine, by filtering them out, or by adding substantial quantities of sulfites. Those people who are against malo argue that drinkers want fresh, zesty wines, whatever it takes. In Germany and Austria, for example, blocked malo is very common.

As far as I'm concerned, blocking malo hampers a wine's development, robbing the drinker of the full flavor and texture profile a wine is capable of. Wines that undergo malo are a lot more expressive than wines that have been purposefully held back and restrained. Allowing malo the freedom to happen is, I think, fundamental to producing natural wine. If the year calls for malo, then so be it. If not, then never mind.

An aside: All the white wines listed here are dry.

Below left: **La Ferme des Sept Lunes' vineyard, in the Rhône, is all about polyculture: the grapes rub shoulders with apricot trees, animals, and grains. They make a range of great wines, including a spicy, white Saint-Joseph, which is worth looking out for.**

Below right: **You can find more information on Hardesty's *Riesling* on page 157.**

FRANCE

FRENCH LIGHT-BODIED WHITES

Recrue des Sens, *Love and Pif*

Hautes Côtes de Nuit, Burgundy

Aligoté

Oyster shell | White pepper | Pear juice

Yann Durieux is one of the most exciting young producers to come out of Burgundy recently. After a 10-year stint working at Prieuré-Roch (which, like Domaine de la Romanée Conti, is another *über*-traditional, *natural* Burgundian estate), Yann broke out on his own. He is definitely a grower to watch and his *Love and Pif*, made from the vastly under-rated aligoté grape, will get you wondering why the so-called *noble* grape varieties have trumped all others as they have done... A wine with surprising depth and detail.

*No added sulfites

Domaine Julien Meyer, *Nature*

Alsace

Sylvaner, pinot blanc

Jasmine | Kiwi | Anis

Although it has many organic and biodynamic farms, Alsace is still reliant on a heavy-handed use of sulfites, which means that growers like Patrick Meyer are few and far between. On taking over the estate, Patrick started eliminating enzymes, yeasts, *et al*, because, as he explained, it just didn't make sense. Today, he is an inspirational grower, with soils so alive they are said to remain warm even in winter. *Nature* is one of the most accessibly priced natural whites: light and fragrant, its texture is almost honeyed, though bone-dry.

*No added sulfites. Filtered

Pierre Boyat, *St-Véran*

Burgundy

Chardonnay

Apple | Sweet hay | Saffron

Pierre is a shy man whose *St-Véran* is exquisite, reflecting the exacting nature of fine natural-wine production. Having run his family estate for decades, Pierre joined a growing cohort of conventional producers who choose to grow and make wine differently. Inspired by Philippe Jambon (a famous low-intervention producer of northern Beaujolais, with whom Pierre now works closely), he sold the domaine and bought up small plots of gamay and chardonnay, which he farms organically. He processes his grapes with minimal intervention, to express the fullness of the *terroir*, and the result is a highly accomplished wine.

*No added sulfites

FRENCH MEDIUM-BODIED WHITES

Andrea Calek, *Le Blanc*
Ardèche, Rhône
Viognier, chardonnay

Fragrant blossom | Stony | Beeswax

This beautiful, quiet, somewhat-forgotten part of the Rhône has over recent years become a hotbed of outstanding natural wine grower-makers (Gilles and Antonin Azzoni, Gerald Oustric, Laurent Fell, Grégory Guillaume, and the Ozil brothers, to name but a few). Maverick Andrea Calek, originally from the Czech Republic, ended up in the wine trade by accident, and thank goodness he did. His wines are profound and uncompromising, showing restraint and complexity. He makes a tiny amount of white wine.

*No added sulfites

Julien Courtois, *Originel*
Sologne, Loire
Menu pineau, romorantin

Smoky | Fresh walnut | Minty

Julien Courtois, one of the sons of the famed Claude Courtois, and his Maori wife, Heidi Kuka—who creates all their beautiful bottle labels—cultivate seven different grape varieties on 11 acres (4.5 hectares) in an area about two hours' drive south of Paris. Julien's wines always show incredible purity, control, and minerality, and the *Originel* is no exception.

*Low levels of sulfites added

Lous Grèzes, *Les Elles*
Languedoc, France
Chardonnay

Woodruff flower | Stone fruit | Beeswax

Lous Grèzes was created in 2002 in the foothills of the Cevennes, a wild part of the Languedoc, by a Belgian couple, Trees and Luc Lybaert. This chardonnay is a beautiful expression of the tight minerality of the limestone plateau where the vines grow. They also make intense, long-lived reds (often released after some bottle ageing) that taste a little like savory, bottled garrigue.

*No added sulfites

Domaine Houillon, *Savagnin Ouillé*
Pupillin, Jura
Savagnin

Fresh walnut | Mustard seed | Acacia flower

Owned and run by the natural wine stalwart Pierre Overnoy for more than three decades, today the domaine is in the very capable hands of Emmanuel Houillon, Pierre's surrogate son. Bottled in June 2012, after eight years in barrel, this savagnin is profound, multilayered, and extremely long on the finish.

*No added sulfites

Matassa, *Vin de Pays des Côtes Catalanes Blanc*
Roussillon
Grenache gris, macabeu

Sage | Toasted almond | Menthol

Prior to setting up shop in Calce, Tom Lubbe's first project—The Observatory—was in the now very trendy South African Swartland. It was a project far ahead of its time, in terms of both its farming and its cellar practices. Matassa follows in these footsteps and, luckily for Tom, it even has a gloriously immense, African-esque view from the top of his Romanissa vineyard. This elegant, light-bodied white, grown on schist, is a medley of dried herbs, with a pinch of salt and a thirst-quenching menthol quality.

*Low levels of sulfites added

Catherine and Gilles Vergé, *L'Ecart*
Burgundy
Chardonnay

Smoky | Honeysuckle | Mineral

The Vergés are possibly the most surprisingly under-the-radar growers I have come across over the last few years. Catherine and Gilles' work is a genuine wonder

Above: **Wine is best stored lying on its side so that the cork is kept moist.**

and their results will bowl over even the most ardent of sulfite-free critics. *L'Ecart* comes from an 89-year-old plot, and the five-year *élevage* is more the rule than the exception. The result is wine that is so stable it can hold its own once open, even for weeks on end. While writing this book, I decided to put a bottle to the test and see how long it could last. I opened it in October 2013 and drank a glass every so often, unceremoniously squeezing the cork back in between sips (oxygen and all) and putting the open bottle back in my humid Victorian coal-chute of a cellar, until mid-January 2014, when I had my last glass. Three months later. It stood up impeccably, even with only a smidgen of wine at the bottom of the bottle. I was aghast.

This uncompromising chardonnay has all the hallmarks of a grand cru. Tense, structured, and extraordinarily fresh, with a steely mouthfeel and bitey minerality, it has great concentration and multilayered aromas, including sweet, fresh butter, and a touch of saltiness and smokiness, alongside heady floral notes. An absolute revelation and a must to drink.

*No added sulfites

FRENCH FULL-BODIED WHITES

Marie & Vincent Tricot, *Escargot*

Auvergne

Chardonnay

Honeydew melon | Mineral | Waxy

Vines arrived in the Auvergne with Caesar in around 50BC and thrived until the beginning of the 20th century when phylloxera struck and decimated the vineyard area. In recent years, however, the Auvergne wine scene has been making a comeback and is now home to a tremendous array of natural wines. I would even go so far as to say that it could well boast one of the highest concentration of natural wine producers anywhere. Blessed with great *terroir* (essentially volcanic soil), natural wines from the region are pure, with a great mineral bite. *Escargot* would not be out of place alongside many Burgundian crus, but at a fraction of the price. Other growers worth keeping an eye out for are: Patrick Bouju, Maupertuis, Le Picatier, François Dhumes, and Vincent Marie.

*No added sulfites

Le Petit Domaine de Gimios, *Muscat Sec des Roumanis*

St-Jean de Minervois, Languedoc

Muscat

Dried rose petals | Lychee | Thyme

Anne-Marie Lavaysse and her son Pierre make some of the purest dry muscats around, from vines grown on limestone outcrops amid wild garrigue. Local growers go for sweet, fortified wines, but Anne-Marie favors dry, and the results, which are produced in minuscule quantities, are beautiful. Compelling and intense, Le Petit Domaine de Gimios' *Muscat Sec des Roumanis* is rich, scented, aromatic, and phenolic. (See *Medicinal Vineyard Plants*, pages 52–53, for more on the Lavaysses' story.)

*No added sulfites

Domaine Etienne & Sébastien Riffault, *Auksinis*

Sancerre, Loire

Sauvignon blanc

Rosemary | Verbena | Smoky asparagus

Unlike any Sancerre you will have tasted before, but by far the best. Sébastien's wines redefine what we think sauvignon blanc should taste like. One of the standard-bearers for the grape, Sébastien makes some of the most profound expressions of sauvignon blanc that exist—a far cry from the zippy, standardized clones that are so much Sancerre today. A meditative, opulent wine, it also has an underlying mineral nerve that roots it firmly in the chalky hills of Sancerre.

*No added sulfites

Domaine Léon Barral, *Vin de Pays de l'Hérault*

Languedoc

Terret with a little viognier and roussanne

White peach | Peppery | Lemon peel

Named after his grandfather, Didier's domaine in the Faugères appellation is no small feat and a role model in polyculture. With 74 acres (30 hectares) of vines, Didier also has another 30 hectares of fields, pastureland, fallow land, and woods, as well as cows, pigs, and horses. Oily in texture and weighty in body, this vintage is particularly perfumed and lifted. His reds, especially the *Jadis* and *Valinière*, have great ageability. (See *Observation*, pages 112–13, for more on Didier's story.)

*No added sulfites

Domaine Alexandre Bain, *Mademoiselle M*

Pouilly-Fumé, Loire

Sauvignon blanc

Acacia honey | Hint of smoke | Salt

Although not technically a "Pouilly-Fumé," as Alexandre recently lost the right to the appellation because of his so-called "atypical" wines (an absurd suggestion, given that he is perhaps the only grower to have a real link between his *terroir* and his final bottle—see page 110 for more details), I have nonetheless chosen to include this wine, as it is, for me, one of the best Pouilly-Fumés around. What is certain is that Alexandre is an oddball in this famous appellation: not only does he farm organically and plow by horse, but also, by refusing to add yeasts and by keeping sulfites at bay, his wines are definitely the finest, most exciting in Pouilly-Fumé today. Luscious and inviting, his *Mademoiselle M* is only one of a raft of sauvignon blanc cuvées, all of which are worth tracking down.

*No added sulfites

Le Casot des Mailloles, *Le Blanc*

Banyuls, Roussillon

Grenache blanc, grenache gris

Almond blossom | Brine | Honey

Founded by iconic natural wine producers Alain Castex and Ghislaine Magnier, Le Casot is today in the hands of Alain's young apprentice, Jordi Perez, who now flies solo. Le Casot crafts a family of sulfite-free wines in a garage (in Banyuls on the Spanish border), using grapes harvested from the schistous terraces lining the walls of the valleys that slice their way from the Mediterranean into the Pyrenees. A storm in a glass, *Le Blanc* is a beautiful wine that is not only stunningly complex, but becomes more linear and restrained with age.

*No added sulfites

ITALY

ITALIAN LIGHT-BODIED WHITES

Cascina degli Ulivi,
Semplicemente Bellotti Bianco
Piedmont
Cortese

Greengage | Aniseed | Citrus

The natural wine legend Stefano Bellotti (who got into trouble with the Italian authorities for planting peach trees in his vineyards—see page 110) has sadly passed away since this book was published, but his work and farm live on. Stefano's focus with his *Semplicemente* trio of wines was always on drinkability. This simple, delicious cortese is no exception. Light, fragrant, and easy-drinking, all of Cascina degli Ulivi's wines are free of sulfites, even their more complex cuvées.

*No added sulfites. Filtered

Valli Unite, *Ciapè*
Piedmont
Cortese

Almond | Fennel | Melon

Valli Unite is a hilltop community in Piedmont. Set up in 1981 by three young farmers, this organic cooperative aimed to buck the trend of rural desertion and enabled them, and others, to remain on the land and follow an alternative lifestyle. They were joined by others, and the result is a community of 35 people who call Valli Unite home. They care for 247 acres (100 hectares) of land and forests, where they farm vines, cereals, chickens, pigs, bees, and vegetables, and run a restaurant and B&B for visitors. Wine is the community's main source of income and they make a great range, including this *Ciapè* and some exciting cuvées of timorasso, an indigenous grape.

*No added sulfites

ITALIAN MEDIUM-BODIED WHITES

Daniele Piccinin, *Bianco dei Muni*
Veneto
Chardonnay, durella

Golden apple | Flint | Honeysuckle

Daniele, Camilla, and their daughter, Lavinia, live in the Alpone Valley, northeast of Verona, where Daniele focuses much of his energy on the growing of the indigenous durella grape. Daniele distills his own herbs, cocktails of which he uses to boost the defenses of his vines (see *Oils & Tinctures*, pages 76–77). This newly released vintage is perhaps the most gentle and inviting of his *Bianco dei Muni* to date.

*Low levels of sulfites added

Nino Barraco, *Vignammare*
Marsala
Grillo

Seaweed | Kumquat | Iodic

Unusually, Nino makes still, unfortified wines in the Marsala region (Marsala is known primarily for its fortified wines). This *Vignammare* is literally planted on a sand dune and was intended to capture "sea in a glass." Although no sulfites are added to the *Vignammare*, the totals of Nino's other wines are typically around 20–35 mg/L. Another of his cuvées worth looking out for is his special *Alto Grado 2009*—an "old-fashioned" Marsala made by late-picking old grillo, which then spends six years under flor in a large chestnut barrel.

*No added sulfites

Orsi Vigneto San Vito, *Posca Bianca*

Emilia Romagna, Italy

Pignoletto, alionza, malvasia, albana, riesling, sauvignon, chardonnay

Fresh walnut | Brine | Grapefruit zest

Federico Orsi and Carola Pallavicino took over this historic farm, located on the hills outside Bologna, in 2005. Alongside their vines, they grow vegetables for local restaurants and also breed Mora Romagnola pigs, who are free to roam (Federico makes a mean "nose-to-tail" mortadella.)

Posca Bianca is the delicious result of fractional blending from a perpetual vat started in 2011. Wines from different vintages, grape varieties, and maturation vessels are all thrown in together, and each vintage fresh wine is added to the mix. The vat is constantly topped up to prevent oxidation. The result is a blend of all their vintages dating back to 2011, or as Federico puts it: "[its] a constantly evolving wine, the synthesis of our *terroir*."

*Low levels of sulfites added

Francesco Guccione, *T*

Sicily, Italy

Trebbiano

Almond | Aniseed | Ripe lemon zest

Unassuming and shy, Francesco biodynamically farms 15 acres (6 hectares) on iron-rich terre-brune soil off the beaten track near Palermo, at about 1,600ft (500m) elevation. His trebbiano grows on pergolas, and although this grape variety may seem out of place on a southern island, given that it is mostly associated with Tuscany and Abruzzo, it has actually been grown in Cerasa since 1400. Francesco's *T* is aromatic, a dash tannic (the grapes spend two days on their skins), and deeply layered. His wines are generally long-lived and benefit from laying down. In his winery, it is not uncommon to see wines still in vats that are five or six years old.

*Low levels of sulfites added

Le Coste, *Bianco*

Lazio

Mostly procanico, with malvasia di candia, malvasia puntinata, vermentino, greco antico, ansonica, verdello, and roscetto

Quince | Nutty | Mineral (volcanic soil)

In 2004, Gian-Marco Antonuzi bought 7 acres (3 hectares) of abandoned hillside in the province of Viterbo, 93 miles (150km) from Rome, on the border with Tuscany, known locally as Le Coste. The name stuck, but the domaine expanded, and today includes olive groves, fruit trees, 40-plus-year-old (rented) vines, and ancient terraces that Gian-Marco and his wife, Clementine Bouveron, plan to use to rear animals. A procanico-dominated blend (85%), Le Coste's *Bianco* ferments for about one year in a *foudre*, in which it stays for another year before bottling.

*No added sulfites

Lammidia, *Anfora Bianco*

Abruzzo

Trebbiano

Salty | White pepper | Almond

Davide and Marco's wines are precisely what they say on the tin: "100% *uve e basta*." Friends since the age of three, these adventurous, young Abruzzese launched themselves into wine production with gusto, naming themselves after the Abruzzese for "evil eye" (*la 'mmidia*). As they explained, "The wise old women of our region cast off envy and the evil eye using an ancient practice made up of water, oil, and magic. After our first harvest, our fermentation suddenly stopped, so we enlisted the help of nonna Antonia who performed the ritual. The fermentation miraculously returned and now nonna takes away *la 'mmidia* before each harvest." Their *Anfora Bianco* spends 24 hours macerating on its skins and then a year in amphora.

*No added sulfites

REST OF EUROPE

REST OF EUROPE LIGHT-BODIED WHITES

Francuska Vinarija, *Istina*

Timok, Serbia

Riesling

Bay leaf | White pear | Dash of lime

"All the best *terroirs* in France have been discovered," says soil specialist Cyrille Bongiraud (who, in a former life, consulted for some 200 different viticultural domaines all over France, including the likes of Comtes Lafon and Zind-Humbrecht, as well as in Italy, Spain, and the United States). This is why he and his vigneronne wife, Estelle (whose great-aunt was Mother Superior at Burgundy's Hospices de Beaune!), spent years looking for the perfect site elsewhere in Europe. The Burgundian couple found it in the chalk valleys of the Danube, in Serbia. Restrained and mineral, the refined *Istina*, with its distinctive petrol notes, is typically riesling, but with the full roundedness you'd expect from a natural wine. (Please note: this wine expressed itself best a couple of days after opening.)

*Low levels of sulfites added

Stefan Vetter, *Sylvaner, CK*

Franken, Germany

Sylvaner

Celery stick | Kaffir lime | Creamy

In 2010, a 60-year-old vineyard in northern Bavaria caught Stefan's eye and, as he puts it: "It was love at first sight." Keen to work with sylvaner, the traditional Franconian grape variety, Stefan now works 4 acres (1.5 hectares) of vines (including a little riesling) and produces wines like this *Sylvaner, CK*, which, while very closed initially, opened over time to reveal a gorgeously delicate, subtle fragrance.

*Low levels of sulfites added

REST OF EUROPE MEDIUM-BODIED WHITES

Strekov 1075, *Rizling*

Južnoslovenská, Slovakia

Rizling vlašský (aka welschriesling)

Red apple skin | Fennel seed | Quince cheese

Zsolt Sütó named his farm after the date of the first record of winemaking in the village of Strekov, a traditional grape-growing part of Slovakia. A creative, innovative grower, his wines are always soulful and textured, and surprisingly both generous and tight at the same time. They combine the mouthfeel and openness typical of so many sulfite-free wines with a crunchy texture, which is a real achievement. He stopped using sulfites across the board in 2017 and does not fine or filter, so his wines often end up cloudy, something which has been an issue for the Slovakian wine authorities, making export tricky. Matured in old oak, Zsolt's welschriesling has a delicious, sweet opulence backed up by great minerality.

*No added sulfites

Gut Oggau, *Theodora*

Burgenland, Austria

Grüner veltliner, welschriesling

Custard apple | White pepper | Cardamom

In 2007, Stephanie and Eduard Tscheppe-Eselböck took over an old property in Oggau with a long tradition of making wine—in fact, some of the walls of this former "Vineyard Wimmer" date back to the 17th century. Apart from making wines with delicious structure, Stephanie and Eduard's genius was to create a multi-generational family of wines, with each cuvée being given its own face and background story to match its personality. *Theodora* started out as one of the youngest in the family, but this easy-going wine becomes more mature with each vintage.

*No added sulfites

Mendall, *Abeurador*

Terra Alta, Spain

Macabeo

Mirabelle | Star anise | Mustard seed

Laureano Serres, the owner of Mendall, a vineyard in the province of Tarragona, is a rarity in Spain, as growers making wines without any added sulfites are few and far between. After working in IT, he did a career U-turn and headed for the outdoors, first as the head of a local cooperative winery (from which he got the sack for trying to help them go at it more naturally) and then on his own. And thank goodness. Laureano makes some of the most stunning sulfite-free Spanish wines—as he says, wine should be "vegetal water or else it is soup."

*No added sulfites

2Naturkinder, *Fledermaus*

Franken, Germany

Müller-thurgau, sylvaner

Sweet pea flower | Earthy | Crushed rice

After discovering natural wine in London, Melanie and Michael quit their jobs and joined Michael's parents at their vineyard in Germany. There they farm traditional Franconian grape varieties (sylvaner, bacchus, müller-thurgau, and schwarzriesling). The vineyards that produce *Fledermaus* (which means "bat") contain a cottage that M & M donated to bat conservation. So was born a symbiotic relationship: they dot bat boxes around the vineyards for their fuzzy friends to hang out in and, in return, the colonies give them guano—a wonderful fertilizer! A share of the profits also goes back to the bats (via the *Landesbund für Vogelschutz*) and, as M & M explain, their label features "the gray, long-eared bat, which has become very rare in our area. It's also incredibly cute and we want to help it stay around."

*No added sulfites

REST OF EUROPE FULL-BODIED WHITES

Weingut Sepp Muster, *Sgaminegg*

Südsteiermark, Austria

Sauvignon blanc, chardonnay

Greengage | Saffron spice | Fresh chestnut

This estate, dating back to 1727, was farmed by Sepp's parents and handed to Sepp and his wife, Maria, on their return home after years spent traveling. Open-minded and avant-garde, this couple have been very progressive in their work in both the vineyard and in the cellar. Maria has two brothers, Ewald and Andreas Tscheppe (see page 156), who also grow natural wine nearby, making for a formidable sibling wine trio in southern Austria. The Muster wines are classified by plot, and *Sgaminegg* (an all-rock *terroir*) is the stoniest, rockiest of their collection, giving the wine elegance and poise.

*No added sulfites

Roland Tauss, *Honig*

Südsteiermark, Austria

Sauvignon blanc

Guava | Passionfruit | Fresh cilantro (coriander)

Roland's natural philosophy embraces all aspects of his life, including the B&B he runs with his wife, Alice, with its breakfasts of his own freshly pressed grape juice and organic honey from their neighbor. Roland is getting rid of everything in his cellar that is not natural—including cement and stainless steel. As Roland explained to me in December 2013, trees take years to grow and have an amazing energy that is imbibed by the wine when barrels are used, whereas vessels made from stainless steel and other cold materials draw energy and strength from the wine. No sulfites had been added to the barrel sample I tasted on my visit (and Roland does not plan to bottle with any either). The wine was still on its lees, and very aromatic, almost gewürztraminer-like, showing exotic fruit aromas. A beautifully pure wine that almost sang.

*No added sulfites

Weingut Werlitsch, *Ex-Vero II*

Südsteiermark, Austria

Sauvignon blanc, chardonnay (or morillon as it is known locally)

Sharon fruit | Flint | Young walnut

Ewald Tscheppe, one of Maria Muster's brothers (see previous page), is particularly interested in soil compaction and soil life. We wandered through his vineyards, as he tried to teach me to read soils by touching the earth and looking at the root structures of various plants on the plots. By digging up some earth you can clearly see where soil is thriving and where it isn't, even in adjacent fields. You notice clear differences in temperature (which soil life helps regulate, so that it is cooler in summer and warmer in winter), in color (a rich soil life means darker soil), and texture (healthy soil is fluffier than its dead, compacted counterpart, which feels more like cement). (For more on healthy soils, see *The Vineyard*: *Living Soils*, pages 25–28.) Born in the beautiful South Styria region of Austria, this wine has a flinty nose, balanced oak spice, and notes of freshly peeled, young walnuts. Bright acidity, great concentration and tension on the palate, suggesting a good few years more of potential development. Although not yet bottled, Ewald assured me this would be bottled without added sulfites.

*No added sulfites

Rudolf & Rita Trossen, *Schiefergold Riesling Pur'us*

Mosel, Germany

Riesling

Ginger | Smoky minerality | Chestnut honey

Organic since 1978, the Trossens definitely go against the (German) grain in what is a pretty conservative home market. In consequence, most of their wines sell abroad. Primarily growers of riesling, which they cultivate on gray and blue slate, the Trossens bottled their first natural wine (nothing added, nothing removed) in 2010, which they then discovered developed completely differently to their other sulfited wines, revealing hidden depths and finesse. They have not looked back since, and so was born their *Pur'us* range. All the wines in the series are really very good, but the *Schiefergold Riesling* (from 100-year-old, ungrafted vines on massively steep slopes) is particularly notable—its concentration, complexity, and length are genuinely extraordinary.

*No added sulfites

UPPA Winery, *Chernaya River Valley, Cler Nummulite Riesling*

Sevastopol, Crimea

Riesling

Waxy | Loquat | Golden gage plum

Ukrainian Pavel Shvets learned all about wine on the floor as a sommelier in Russia. He even won the title of "Best Sommelier in Russia" in 2000. After returning to Sevastopol in the noughties, he created the UPPA Winery in 2008, pioneering biodynamic viticulture in the region. Sadly, being located in the Crimea means that UPPA has been caught up in the current political turmoil, making it difficult for Pavel to export his wines. You may find them difficult to find too, particularly if you're based in Europe. This riesling spent 18 months on the lees and has a dash of fizz on opening. Pavel makes dozens of other wines, including a great range of pet nats.

*No added sulfites

Georgas Family, *Black Label Retsina*

Attica, Greece

Savatiano

Lime peel | Pine needle | Saline

Dimitris Georgas is a fourth-generation grower who took over the family vineyards outside Athens in 1998, and converted them to organic farming. He produces wine, of course, but also grape juice, traditional grape concentrate, grape water, and one of the best retsinas out there. This vibrant, pale lemon wine is made with the traditional savatiano grape, macerated on its skins for a week and fermented with local Aleppo pine resin—an ancient Greek custom dating back thousands of years (circa 1700BC) when resin was used as a preservative. The practice was bastardized in the 1960s and '70s due to the sale of poor-quality retsinas in Athenian tavernas. This tarnished the drink's reputation and left a lingering association with cheap vacation tipples. Dimitris' retsina was a revelation. Highly refreshing, it is utterly delicious.

*No added sulfites

NEW WORLD

NEW WORLD LIGHT-BODIED WHITES

Les Pervenches, *Le Couchant*

Quebec, Canada

Chardonnay

Curry leaf | Mace | Golden Russet (apple)

This 17-acre (7-hectare) farm (with 10 acres/4 hectares under vine) was established in 1991 and taken over by Véronique Hupin and Michael Marler in 2000. They converted the farm to organics and then biodynamics a few years later. *Le Couchant* comes from a well-drained, sandy, gravelly plot of chardonnay planted in 1991, making these the oldest chardonnay vines in Quebec. To the north and west is a large maple woodland that protects the vineyard from the strong westerly winds. This helps to create a warm microclimate that is perfect for growing chardonnay.

*No added sulfites

Hardesty, *Riesling*

Willow Creek, California, USA

Riesling

Green lime | Grapefruit | Dry sage

Born in southern California, Chad Hardesty's love of the land saw him head north to work on, and then start up his own, organic fruit and vegetable farm, where he produced for local restaurants and farmers' markets. Then, under the tutelage and guidance of the pioneering Californian wine grower Tony Coturri, Chad moved into wine, with his first commercial vintage appearing in 2008. This is a young grower-winemaker, whose precision crafts mineral-driven wines with restraint and tightness, as much in terms of reds as of whites. His 2010 riesling is moreishly fresh and steely. I am also a big fan of his *Blanc du Nord*. A guy to watch.

*No added sulfites

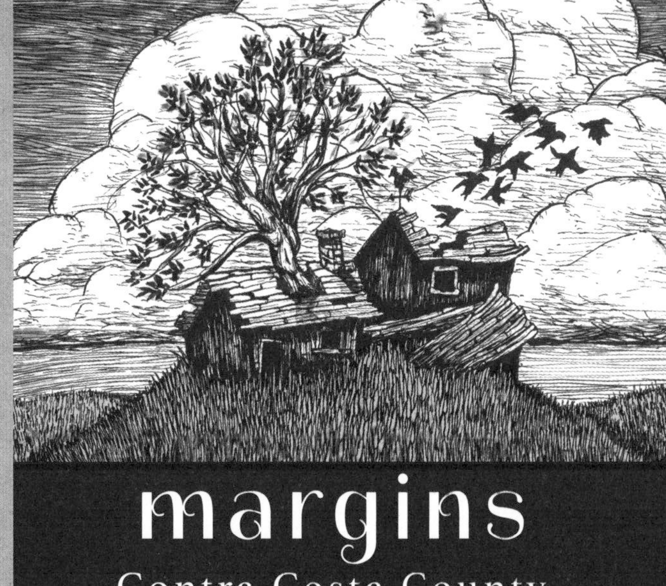

Above: **Megan Bell, of Margins Wine, works with vineyards in the Santa Cruz mountains, where she is part of a dynamic new guard of young Californian makers who take an active part in the farming of the grapes they buy in, with the aim of pushing the organic agenda.**

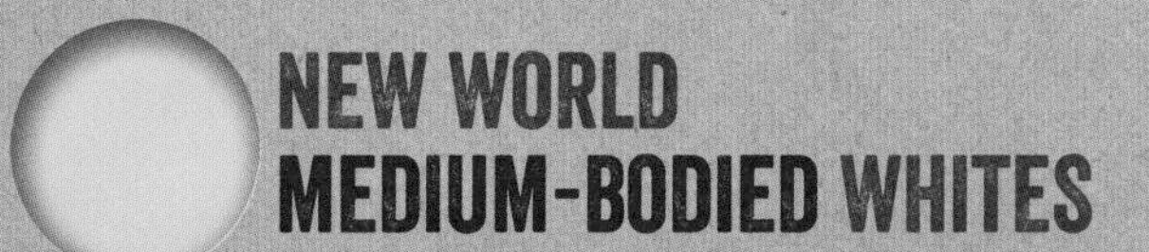

NEW WORLD MEDIUM-BODIED WHITES

Stirm Wine Company, *Wirz Vineyard, Riesling*

California, USA

Riesling

Pink grapefruit | Chamomile | White blossoms

Given the price of land locally, Ryan Stirm supplements the grapes he buys in by leasing abandoned, remote, or diseased vineyards that he nurtures back to health, almost like a vine doctor. Ryan's grape of predilection is riesling. This richly flavored, mineral-driven wine hails from low-yielding, dry-farmed old vines, planted in 1964. His winery is also an incubator for many other talented young makers, such as Megan Bell (Margins Wine, see above) and James Jelks (Florèz Wines), who typically swap help in the vineyard for space in the winery.

*Low levels of sulfites added

Above: **A line-up of multiple vintages of late-harvest and overnight pressings of riesling and gewurztraminer, prior to blending, at Bloomer Creek Vineyard, in the Finger Lakes, USA.**

Bloomer Creek, *Barrow Vineyard*

Finger Lakes, USA

Riesling

Wild peach | Citrus | Dried apricot

For Kim Engle and his wife, Debra Bermingham, wine is a form of artistic expression capable of concentrating experience and memory. Their Bloomer Creek vineyard has taken 30 years to create. It is farmed with care, by hand, and features very slow fermentations in the cellar—often the wines' malo only completes the summer after harvest. Given the cool climate of the surrounding area, I was expecting something a lot more austere when I first tasted this riesling. Instead, I discovered a suppleness and generosity that surprised me (thanks to the Lakes' temperate microclimate). I particularly loved the contrast of its gentle, almost creamy, texture with its biting minerality and great complexity.

*No added sulfites

Sato Wines, *Riesling*

Central Otago, New Zealand

Riesling

Honeysuckle | Nectarine | Allspice

At 45 years old, investment banker Yoshiaki Sato and his wife, Kyoko, ditched their city jobs and plunged themselves headfirst into the world of making wine, eventually setting up shop in Central Otago, in New Zealand. To hone their skills, they worked vintages in both the southern and northern hemispheres, including, among others, chez the revered Pierre Frick in Alsace. The Satos' wines are beautifully crafted, so keep your eyes peeled for their pinot noirs too.

*Low levels of sulfites added

Si Vintners, *White SI*

Margaret River, Australia

Sémillon, chardonnay

Green mango | Baked apple | Gingery

Sarah Morris and Iwo Jakimowicz (SI) spent a few years working in a cooperative in the province of Zaragoza, in Spain, before deciding in 2010 to head home and settle in Margaret River where they bought a 30-acre (12-hectare) estate (20 acres of which are under vine). Loathe to relinquish Spain completely, however, they started up a Spanish label with friends, and so divide their time between Spain and Australia. Fermented in a mixture of concrete eggs, oak *foudres*, and stainless steel, 120-dozen bottles of this wine were made. (Keep your eyes peeled for Paco & Co, their Calatayud project based on 80-plus-year-old grenache vines.)

*Low levels of sulfites added

La Clarine, *Jambalaia Blanc*

Sierra Foothills, California, USA

Viognier, marsanne, albariño, petit manseng

Apricot | Ripe melon | Hay

Inspired by the writings of Masanobu Fukuoka (see *The Vineyard*: *Natural Farming*, page 36), Hank Beckmeyer began to question the foundations of farming, wondering what would happen if he gave up trying to control outcomes (even organically), by taking on the role of caretaker, rather than active participant. "Dropping what is comfortable and 'known' and deciding to... trust in natural processes may seem very daring... But," as Hank explains on his website, what it really requires is "commitment... [and] an acceptance of chance [that] also allows for the possibility of failure."

Today, his 10-acre (4-hectare) farm grows grapes, raises goats, and is shared with countless dogs, cats, bees, chickens, birds, gophers, flowers, and herbs. This generous, mouth-filling, and yet refreshing tipple is very Rhône-esque in style (as are many of his wines, in fact). Hank also makes delicious reds that are worth tracking down, particularly the Sumu Kaw syrah from a vineyard at 3,000ft (900m) of altitude.

*Low levels of sulfites added

Populis, *Populis White*
Northern California, USA
Chardonnay, colombard

Greengage | Citrus | Pear

Many young, low-intervention winemakers in the US buy in grapes. This can sometimes be confusing, as they end up with one organic cuvée and another that uses grapes farmed conventionally. This isn't the case with Populis. They do buy in grapes, granted, but only old-vine, organic ones from northern California. Diego Roig, Sam Baron, and Shaunt Oungoulian created Populis as a label for their family, friends, and allies—a wine for the people—that was clean, delicious, and affordable, as they realized there simply wasn't enough reasonably priced wine with soul locally to go round. No additions or interventions are used in the cellar and the result is great fermented grape juice.

*Low levels of sulfites added

Hiyu Farm, *Falcon Box*
Columbia Gorge, Oregon, USA
Pinot noir, pinot gris, pinot blanc, aligoté, chardonnay, pinot meunier, melon de Bourgogne

White peach blossom | Jasmine | Earthy

Hiyu is a 30-acre (12-hectare) farm in the Hood River Valley. Here permaculture meets biodynamics, and a holistic view of agriculture prevails. There is no mowing or tilling on site; instead, the pigs, cows, chickens, ducks, and geese that live among the vines control the vegetation. Vines rub shoulders with vegetable gardens, woodland, and pastures, and even the vineyards themselves are biodiverse and eclectic, with over 80 different grape varieties co-planted in small blocks and harvested together, reflecting founder Nate Ready's vision of *terroir*. This field blend of Burgundian varieties shows tremendously crystalline aromas.

*Low levels of sulfites added

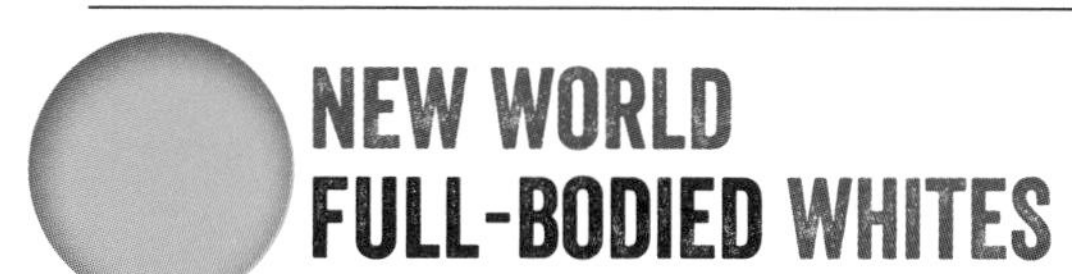

NEW WORLD FULL-BODIED WHITES

AmByth, *Priscus*
Paso Robles, California, USA
Grenache blanc, roussanne, marsanne, viognier

White peach | Licorice stick | Sweet pea

Welshman Phillip Hart and his Californian wife, Mary Morwood Hart, dry-farm their entire estate in Paso Robles (see *Dry-Farming*, pages 38–39)—no mean feat (and very admirable), given California's water wars and when you know that in 2013 they got a grand total of half an inch of rainfall! They also work without any added sulfites (again a rare feat in California). Their *Priscus*, meaning "venerable and ancient" in Latin, is a wholesome, herbal wine that is utterly delicious, as is all their range.

*No added sulfites

Dominio Vicari, *Malvasia da Cândia e Petit Manseng*
Santa Catarina, Brazil
Malvasia di candia, petit manseng

Lime peel | Passionfruit | Asparagus

Created in 2008 by Lizete Vicari (a ceramist) and her son, José Augusto Fasolo (an oenologist), in their garage, Dominio Vicari is today one of Brazil's cult artisan producers and part of a small, but burgeoning,

scene of natural grower-makers in the country. Using grapes grown by their family in Monte Belo do Sul, Rio Grande do Sul, Lizete—who fell in love with the art of low-intervention winemaking—made a name for herself producing orange wines from Riesling Italico grapes, a variety that apparently abounds in Brazil. Today, she and her son make wines from all sorts of varieties, including merlot, cabernet, sauvignon blanc, grechetto, ribolla, and others. All are made naturally, without temperature control, fining, or filtering.

*No added sulfites

Scholium Project, *The Sylphs*
California, USA
Chardonnay

Green mango | Salty | Sweet oak

Named after the Greek word for "comment" or "interpret" (from which we also derive the words "school" and "scholarship"), Abe Schoener created Scholium as a "modest project... for the sake of learning and understanding." The result is a wild bunch of soulful wines created from vines he leases. *The Sylphs* is a dense, textured, oaky wine, but with well-balanced fruit. Another favorite, from my limited tasting of his wines, was a skin-macerated sauvignon blanc: *Prince in his Caves*.

*No added sulfites

Caleb Leisure Wines, *Chiasmus*
Sierra Foothills, California, USA
Marsanne, roussanne, viognier

Barley | Apricots | Lily-of-the-valley

Caleb, a lovely, young Californian native (now also slightly English by marriage), is at the very beginning of his winemaking adventures. This is his very first vintage and what a lovely 35 cases (of viognier/marsanne/roussanne pet nat) and one barrel (of this white) he has created! They are both generous and juicy.

I am particularly thrilled to include Caleb in this third edition of *Natural Wine*, as it really is a question of life coming full circle, since it was thanks to the first edition of this book that Caleb turned up on Tony Coturri's doorstep (book in hand) and declared: "I have come to see you because you were featured in this book." Never having worked in wine before, he struck up a friendship with Tony, started working for him, and now makes his own tiny production in Tony's cellar. Hopefully, this heartwarming story will inspire other lovely encounters in the world of wine too.

*No added sulfites

Coturri, *Chardonnay*
Grebennikoff Vineyards, Sonoma Valley, USA
Chardonnay

Linden (lime) | Grilled hazelnuts | Honey drops

Californian Tony Coturri (see *Apples & Grapes*, pages 128–129) is the natural wine veteran of the USA and it's high time he received the recognition he deserves. An ex-hippy, Tony started his Sonoma farm in the 1960s and has produced dozens of vintages of delicious, organic, no-added-sulfite wines. Isolated, in a world of not-very-like-minded souls, Tony, a farmer at heart, was treated like a bit of a crazy. "Growers round here don't call themselves 'farmers.' They call themselves 'ranchers.' There's a big difference. A farmer, in their minds, is a negative—some guy wearing overalls, trying to eke out a living from chickens and stuff. 'We're ranchers,' they say. The whole thing is so skewed. They don't even see viticulture or grape-growing as an agricultural pursuit," explains Tony. The result is that wines chez Tony are extraordinary. *Terroir*-driven and authentic, his full-bodied cuvées have a depth and quality that are remarkable. As creamy as nectar, 80 cases were made of his chardonnay.

*No added sulfites

LIGHT-BODIED WINES

MEDIUM-BODIED WINES

FULL-BODIED WINES

Have you ever wondered why, in Renaissance paintings, the white wines in people's glasses don't look as translucent as the whites of today, and why they look more orange than anything else? This isn't a trick of the light or the patina of age, but perhaps because the Michelangelos of this world actually drank orange wine.

Today, most white wines are made by pressing the grapes, separating off the juice, and discarding the skins, stems, and pips to produce a wine that is pale in color. If, instead, the juice is left to macerate and ferment on its skins, pips, and possibly stems, you end up with a wine that looks orange—or any of a variety of hues, ranging from yellow to Orangina to Fanta or even shades of rust. The maceration can be as short as a few days or extremely developed, sometimes lasting for months (like the Italian Radikon, on page 169) or even years (as is the case with the South African Testalonga, also on page 169).

ORANGE WINES

While a seemingly new trend, orange wine is ancient. When white wine production first began, it would, most likely, have been made like red wine, using the whole berry rather than the run-off juice, which is much more fiddly and complicated to work with given its vulnerability to oxygen. Indeed, as Dr. Patrick McGovern from the University of Pennsylvania explains, "An Egyptian jar, dating back to 3,150BC, showed yellowish residue, together with seeds and skins, which would seem to point to maceration." Similarly, skin-macerated whites may actually be the "yellow" Pliny refers to when he writes that, "wine has four colors... white, *yellow*, red, and black."

WHERE IS ORANGE WINE MADE?

Although starting to make an appearance in the public's consciousness and fast gaining in popularity, orange wines are still pretty rare when compared with reds, whites, or even rosés. Excellent examples exist in places as diverse as Sicily, Spain, and Switzerland, but the main hotbed of production is undoubtedly Slovenia and the neighboring Italian Collio, where the most profound examples of this style of wine can be found. The Caucasian Georgia probably boasts the largest number in terms of total production, given that many Georgians produce it at home.

Left: **Macerating white grapes in their juice helps extract aromas, structure, and color, which define orange wines.**

Opposite: **Orange wine exists in a panoply of hues, from yellow to luminous orange, or even dark amber.**

Nowadays, orange wine has become very fashionable, which means that there is a lot of jumping on bandwagons. As such, not all wines *called* orange, *are* orange, as the wine has to taste orange too. For Saša Radikon, one of Europe's foremost orange wine producers, this means that the wine "has to be macerated on natural yeasts and with no temperature control. Then, even just five days on the skins, and it can be completely orange. With temperature control—even at just 20°C—you could macerate for a whole month and still not get the color because it is too cold to extract it."

What's more, not all wine producers can actually legally produce orange wines, either because the category is missing from official, formalized wine structures (such as appellation bodies, regulations on export, and the like—see Craig Hawkins' story, pages 108–109) or because it is simply forbidden. Pascal Doquet, President of the *Associations des Champagnes Biologiques*, explains that growers who produce wines in Champagne are obliged by law to produce wines that have the AOC (i.e. that have the appellation). These growers are not allowed to voluntarily desist from the AOC, either for a particular cuvée or indeed for all their wines, and produce a Vin de France instead, an option that is open to most other producers in France. "So you either produce an AOC wine or your wine is sent to the distillery," explains Pascal, and the issue is that the AOC regulations stipulate that "grapes have to be pressed whole, meaning there cannot be any maceration whatsoever prior to pressing. The result is that Champagne or Côteaux Champenois wines have to be white." They cannot be orange.

WHAT DOES ORANGE WINE TASTE LIKE?

Orange wines are likely to be some of the most unusual wines you'll ever taste. Although sometimes confrontational, the best examples are soulful and complex, featuring unexpectedly novel combinations of flavors and textures. The most curious aspect is their tannic intensity. During skin contact, tannins (and antioxidants) are extracted, giving these hybrids the texture of a red. Taste them blind, or in a dark glass, and it is hard to know which side they bat for.

Try orange wines alongside food and they come into their own, even becoming highly addictive. The tannins soften, even disappear, and the versatility of their extraordinary flavor profile becomes apparent. These wines work well with a wide range of dishes, but are particularly suited to robust flavors such as mature hard cheeses, spicy stews, or, even better, walnut-based dishes. They also seem to show their best when drunk out of large glasses, as they need plenty of air in order to open up and reveal their full personalities. The key is to treat orange wines more like reds than whites, and not to drink them too cold.

One of the criticisms sometimes leveled against orange wines is that they all taste the same. It is true that macerating juice and "bits" has an impact on the flavor profile, color, and texture of the wines, but they are nonetheless able to express particularities of *terroir* (be it the volcanic soil of Etna or the decomposed granite from the Swartland), as well as the grape varieties used (be it lean rebula or fleshy, spicy pinot gris). All in all, not very samey.

Did you know? The term "orange" was first used (and perhaps invented) in 2004 by David Harvey, a UK wine-trade professional. "There was no industry standard at the time, even the growers themselves had no name for it", explains David, "and since we use colors for everything else, it seemed obvious because that is what it is, orange."

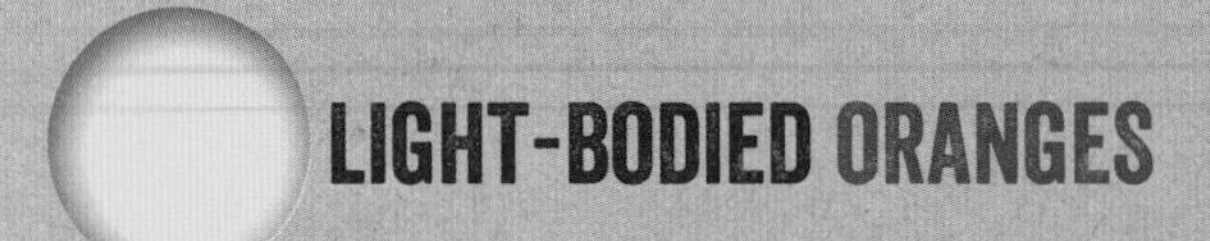

Négondos, *Julep*
Quebec, Canada
Seyval blanc

Yuzu | Celery seed | Yellow nectarine

Imagine having to brave a cold, harsh climate *and* farm organically? Grape-growing in Canada is no small feat. Négondos dates back to 1993 and was the first organic winery in Quebec. They work entirely with hybrids. These are grape varieties that are developed by crossing two or more vine species (i.e. not solely *Vitis vinifera*) with a view to creating plants that are better suited to the climate, able to ripen when other varieties would struggle, or that are resistant to specific diseases, for example. This skin-contact seyval blanc—a French hybrid—is delicate and light, and pretty Nordic in style in that it almost has an icy quality. You can tell it was grown in a very marginal climate as there is a purity and clarity of aromas that is almost crystalline.

*Low levels of sulfites added

Escoda-Sanahuja, *Els Bassots*
Conca di Barbera, Spain
Chenin blanc

Smoked hay | Dry quince | Fenugreek

This Spanish chenin blanc, from a calcareous outcrop in northeastern Spain, is a bit of an oddball. It is both a Spanish chenin (which is wholly unusual) and macerated by Joan-Ramón Escoda on the skins for eight days. The result is a bone-dry, linear wine with a hint of tannins. Only 4,500 bottles are produced.

*No added sulfites

Sextant, *Skin Contact*
Burgundy, France
Aligoté

Honeycomb | Kaffir leaf | Dragon fruit

Julien Altaber is a very promising vigneron who used to work with iconic Burgundy producer Dominique Derain. He is part of an exciting new generation of winemakers who are brave enough to make different wines in extremely classic, conservative wine regions. Aligoté is often regarded as a second-rate grape, known for making insipid, acidic wines, whose only raison d'etre is as a sidekick to crème de cassis in the making of Kir. However, Julien's wine proves that, in the end, farming is everything. In other words, when grapes are farmed properly and wines are made naturally, any grape is capable of great personality. Julien's *Skin Contact* was macerated for 12 days (50% whole bunch; 50% de-stemmed) and shows gentle purity and upfront fruit. It is elegant and very much in the style of all his cuvées.

*Low levels of sulfites added

MEDIUM-BODIED ORANGES

Le Soula, *Macération*
Roussillon, France
Macabeu, vermentino

Citrus peel | Licorice stick | Persimmon

The dynamic South African Wendy Paillé is now at the helm after taking over from Gerald Standley, who put this great domaine on the map. All their wines are beautiful, but they do maceration particularly well. The *Macération*, which goes through two weeks of skin contact, is both perfumed and floral with a soft, plush texture. Many orange wines lack generosity and end up quite heavy, with mouth-drying tannins and a general lack of aromatic freshness. The most successful orange wines are the ones that manage to retain a lot of lifted notes, just like this one.

*Low levels of sulfites added

Denavolo, *Dinavolino*

Emilia-Romagna, Italy

Malvasia di candia aromatica, marsanne, ortrugo, and others

Fresh fennel | Orange rind | Coriander seed

Created by Giulio Armani, resident winemaker of La Stoppa, Denavolo is Giulio's personal 8-acre (3-hectare) project. Macerated for two weeks on skins, *Dinavolino* is a great introduction to orange wines, with enough weight of fruit to balance the tannins, while being quite tight in style. While shy on the nose, it reveals a generously open and aromatic palate.

*No added sulfites

Ökologisches Weingut Schmitt, *Orpheus*

Rheinessen, Germany

Pinot blanc

White peach | Honeysuckle | Burdock

Daniel and Bianca Schmitt farm 37 acres (15 hectares), and are one of only 75 Demeter-certified wineries in Germany. About half of their production is made using skin contact, is unfiltered, and bottled without any sulfites (a rarity in Germany, where most producers, even organic and biodynamic ones, rely heavily on them). In the case of *Orpheus*, the juice spends two months on its skins and is aged in Georgian *qvevri* (or *kvevri*) for a year.

*No added sulfites

Elisabetta Foradori, *Nosiola*

Trentino, Italy

Nosiola

Acacia blossom | Macadamia nuts | Brine

Elisabetta vinifies this wine in Spanish *tinajas*, which are clay jars that she keeps above ground. The grapes are de-stemmed, and then macerated on the skins for six or seven months, after which the wine is transferred to old acacia barrels for three months or so. The result is extraordinarily delicate and fragrant (and a touch flowery), with gentle tannins and a salty edge. It is such a pretty wine that it almost doesn't feel like an orange.

*Low levels of sulfites added

Colombaia, *Bianco Toscana*

Tuscany, Italy

Trebbiano, malvasia

Hazelnuts | Salty caramel | Pears

Dante Lomazzi and his wife Helena's 10-acre (4-hectare) vineyard is cultivated on clay-calcareous soils, which the Lomazzis farm, in their words, as "a big garden." This opulent, salty blend has great density and a tannic edge, thanks to some skin contact. Look for their limited-edition sparkling white and rosé, *Colombaia Ancestrale*.

*Low levels of sulfites added

Mlečnik, *Ana*

Vipavska Dolina, Slovenia

Chardonnay, tocai

Fresh tobacco leaf | Saffron | Hint of peach

Walter Mlečnik makes refined orange wines. Released with at least four to five years of maturation, they show great complexity and maturity. Elegant, with a hint of spice, *Ana* shows unusual restraint for an orange wine.

*Low levels of sulfites added

Fattoria La Maliosa, *Bianco*

Tuscany, Italy

Procanico, greco piccolo, ansonica

Curry leaf | Linden (lime) honey | Lime zest

Antonella Manuli's 407-acre (165-hectare) farm, in the Maremma Hills, is a haven of sustainability. Arable lands rub shoulder with ancient grape varieties, as well as 70-year-old olive trees and forests. The resulting delicious products include natural wines, extra virgin olive oil, and luscious honey. Farmed according to the "Metodo Corino," a set of rules created by celebrated agronomist Lorenzo Corino (see *Case Corini*, page 188), who helps Antonella on her estate, La Maliosa is proud to be certified organic and suitable for vegans, and even tracks its carbon emissions, so committed is it to the environment.

*No added sulfites

Cornelissen, *Munjebel Bianco 7*

Sicily, Italy

Carricante, grecanico, coda di volpe

Kumquat | Green mango | Squirt of lime

The Belgian-turned-Sicilian Frank Cornelissen has had many lives: alpinist, racing-car driver, wine importer, and now grower of fine natural wines on the lava fields of Mount Etna. Made from local Sicilian grapes, his *Munjebel Bianco 7* is unruly and wild, a bit like experimental jazz.

*No added sulfites

FULL-BODIED ORANGES

Gelveri, *Mayoglu Terebinth*

Cappadocia, Turkey

Keten gomlek

Camphor | Walnut | Dried marigold flower

Inspired by *qvevri* (*kvevri*) producers that he met while working in Georgia, Udo Hirsch (the owner of Gelveri) started making natural wine in 2010 near the 10,500-ft (3,200-m) high Hasan Dag volcano, in Turkey. He uses grapes grown on volcanic tuff at an altitude of 5,000ft (1,500m) in traditional private fruit gardens. Some of these vines are over 200 years old and all are ungrafted. Every year, each vine is fertilized with a shovel of goat dung and nothing else.

Mayoglu Terebinth is a beautifully original wine, based on a family recipe given to Udo by the previous owners of his home and cellar. This orange marvel is a skin-macerated keten gomlek (an indigenous Turkish grape variety) that has been infused with the fruit of the terebinth tree, which grows near his home. "The terebinth tree is part of the Pistachio family," explains Udo. "Its fruit, leaves, flowers, and resin are used for all sorts of different purposes, and in ancient times it was shipped to Egypt for the production of perfume and for mummification!"

The first step is to make the wine. Juice, skins, pips, and stems are fermented in an open *küps* (a large clay jar

Above: **Udo Hirsch's Gelveri vineyards in Turkey. The Hasan Dag volcano can be seen in the background.**

that is the Turkish equivalent of the Georgian *kvevri*), which sits above ground. Once fermentation is complete, the *küps* is closed for a minimum of six months, and only once the wine is ready is the terebinth fruit finally added. "I let the fruit steep in the wine for exactly two years," explains Udo. The result is a fascinating wine, probably unlike anything you have ever tasted before. I certainly hope that Udo continues to make it each year.

*No added sulfites

Pheasant's Tears, *Mtsvane*

Kakheti, Georgia

Mtsvane

Chamomile | Pomelo | Almond

There is a saying in Georgia that "only a wine beyond measure could make a pheasant cry tears of joy," and Pheasant's Tears certainly has a good shot at it. They produce a raft of highly enjoyable, traditional Georgian wines from autochtonous grapes (the country boasts hundreds of native grape varieties), using ancient recipes, which sees skins, pips, stems, and juice left for six months in a *qvevri* (or *kvevri*)—a large clay jar buried underground. Floral with wonderfully integrated, delicate tannins, this *mtsvane* is a delicious example of what Pheasant's Tears, and Georgia in general, has to offer.

*Low levels of sulfites added

Čotar, *Vitoska*

Kras, Slovenia

Vitoska

Sweet quince | Licorice | Lemongrass tea

Located 3 miles (5km) from the sea, just north of Italy's Trieste, the father-and-son team Branko and Vasja Čotar make elegant, linear wines. While bone-dry, their *Vitoska* has a perfumed sweetness with some sediments in the bottle, which I like to shake before serving in order to make the most of the fullness of texture.

*Low levels of sulfites added

Cantina Giardino, *Gaia*

Campania, Italy

Fiano

Smoked hay | Mandarin | Passionfruit

With only four days on its skins, this is really a borderline orange/white, since little extraction will have taken place in this short space of time. Yet this short period on the skins has clearly marked the wine's texture (check out its touch of tannins, for example) and aromatics. *Gaia* hails from old fiano vines grown on volcanic soil, in Irpinia, high in the hills of Campania. Giardino's wines brim with life and are always extremely vibrant and fresh.

*No added sulfites

Above: ***Qvevris* (or *kvevris*) at Pheasant's Tears (opposite), lined with beeswax and ready to be interred. These large clay jars are used for wine fermentation and maturation throughout Georgia.**

Serragghia, *Zibibbo*

Pantelleria, Italy

Zibibbo

Ylang ylang | Passionfruit | Sea salt

Gabrio Bini makes some extremely individual wines on this volcanic mass close to the African continent. Horse-plowed, these zibibbo (aka muscat of Alexandria) grapes are left to ferment in old amphorae buried outside. Gabrio also grows and preserves some of the best natural capers you will ever taste. This bright, luminous orange wine is an explosion of intense exotic flavors mixed with a strong sea breeze.

*No added sulfites

Testalonga, *El Bandito*

Swartland, South Africa

Chenin blanc

Fresh hay | Apricot | Dried apple peel

Made by Craig Hawkins, a daring South African producer who makes wine for Lammershoek in the Swartland, Testalonga is Craig's solo venture. This uncompromising, dry-farmed chenin blanc was kept in oak, on its skins, for two years. Deliciously easy-drinking, *El Bandito* is surprisingly juicy for a wine that shouldn't be on paper. With warm spices, which are inviting, and a refreshing acidity, *El Bandito* is a hugely moreish wine.

*No added sulfites

Radikon, *Ribolla Gialla*

Friuli, Oslavje, Italy

Ribolla gialla

Marmalade | Star anise | Almond

One of the most thrilling orange wines out there, Radikon's *Ribolla Gialla* is a journey in a glass, as it changes immensely between first pouring and one hour down the line. The result of extended skin contact, lasting several weeks, and lengthy ageing (over three years) in large oak barrels, this is a wine of brutal complexity, and a depth of flavor that is both cold and bold, almost stoic in character.

*No added sulfites

LIGHT-BODIED WINES

MEDIUM-BODIED WINES

FULL-BODIED WINES

Called *rosé*, *blush*, or *vin gris* ("gray" wine), the last two being names for very pale examples of the type, pinks are wines that are essentially made using grape varieties with red skins—and sometimes red flesh—whose color leaches into the wine. The result of this contact can be any shade of pink (or pale purple), from onion-skin to salmon or copper, bright pink or even deep fuchsia. In fact, at the darkest end of the scale, some pinks can almost look red.

How dark or light a rosé becomes depends on a variety of factors, including the length of time the juice sits on the skins or the pigment strength of the grape variety used. Different grapes impart more or less color, as well as more or less structure, to the wine. It would, for example, be more difficult to obtain a deeply colored, full-bodied rosé using a thin-skinned grape variety like pinot noir than it would if you used a thick-skinned, anthocyanin-rich cabernet sauvignon. Equally, it would be hard to make a light rosé from *teinturier* grape varieties, such as alicante bouschet or saperavi, with their deep black skins and colored flesh.

PINK WINES

The color and structure of pinks are, however, also down to the recipe used for the wine's production, which might, for example, see growers:

- Blending red and white wines together, as is the norm for much pink Champagne production (in fact, in France, according to AOC rules, *only* Champagne can employ this method);
- Doing short skin macerations on their reds;
- Using what is known as the "*Saignée* Method," when the top of a vat is "bled" by skimming off part of the red-wine production at the very beginning of the fermentation process, so giving the grower a rosé, as well as a more concentrated red;
- Or, even lower down the (conventional) quality chain, using additives or processing agents, such as activated carbon, to strip a red wine of its color.

The quality of the results certainly varies, especially since producing a rosé is often, unfortunately, an afterthought of the winemaker. Indeed, winemakers will often focus on whites and reds, using the best berries for the endeavor, after

which they use the leftovers to spit out a rosé—which consequently becomes a sort of by-product of the red-wine production. As a result, many rosés end up being wishy-washy examples that don't seem to have made up their minds where they stand, falling somewhere between a white and a red.

The most important factor, therefore, is the intention behind the production, as the best rosé examples are undoubtedly those that are made from grapes specifically intended for rosé.

In recent years, the popularity of pinks has sky-rocketed, which seems to have paved the way for volume-driven brands to line store shelves with insipid, off-dry, hard-candy-like examples, sometimes putting discerning eaters or drinkers entirely off the category as a whole, which is a shame. For those of you in that camp, the selection that follows should prove a pink rebirth. They are beautiful (dry) wines in their own right, and one of them is, in fact, a serious contender for being my personal Desert Island wine...

Above: **Pierre Rousse, of Le Pelut vineyard, in the Languedoc, France, makes a raft of sulfite-free wines, including a pinot-noir rosé called *Fioriture*.**

Above: **Meet Cali the bull. He is one of a herd of 12 Highland cattle that overwinters among the vines of Mamaruta (another vineyard in the Languedoc), helping to fertilize the soil in the process. Look out for Mamaruta's *Un Grain de Folie*, which is made with a dash of sulfites at bottling.**

LIGHT-BODIED PINKS

Mas Nicot

Languedoc, France

Grenache, syrah

Wild strawberry | Raspberry | Undertone of cacao

Husband-and-wife team Frédéric Porro and Stéphanie Ponson make some of the most affordable and reliable natural wines out there. This rosé blend of grenache and syrah has the clarity of soft, red fruits, with a slightly darker, more serious edge—a hint of spice and tannin. Look out for their other estates as well: Mas des Agrunelles and La Marele.

*Low levels of sulfites added

Domaine Fond Cyprès, *Premier Jus Rosé*

Languedoc, France

Carignan, grenache

Ginger | Rhubarb | Pink lady

After meeting famed Burgundy producer Fred Cossard in 2004, Laetitia Ourliac and Rodolphe Gianesini (owners of Fond Cyprès) decided to make natural wine. As they explained, "We have worked hard all these years to find the vineyard's identity: to know each plot, to try different wine-producing methods, to taste our close and far neighbors. It is how we got here." The result of this labor of love is a series of wines of surprising honesty and a lot of personality—this carignan/grenache blend, for example, is very gregarious and aching to be drunk in large quantities. Look out also for their *Le Blanc des Garennes*, which is the vinification of a single plot where grenache blanc, roussanne, and viognier are co-planted.

*No added sulfites

MEDIUM-BODIED PINKS

Franco Terpin, *Quinto Quarto, Pinot Grigio delle Venezie IGT*

Friuli, Italy

Pinot grigio

Blood orange | Fennel | Wild raspberry

Although he is better known for his concentrated, serious, orange wines, Franco's spicy, gregarious pinot grigio is highly quaffable. Almost zesty, with some savory, aniseedy notes that help with balance, it is a fun, happy sort of wine. Bouncy and vibrant, it is hugely expressive, juicy, and refreshing.

*Low levels of sulfites added

Mas Zenitude, *Roze*

Languedoc, France

Grenache, cinsault, carignan

Red plum | Bay leaf | Vanilla

This wine by Swedish lawyer-cum-wine-producer Erik Gabrielson is surprisingly serious. Lacking the energy of, for example, Franco's wine above, it is far more subdued and heavy: rounder, spicier, and richer. It is textured and has a creamy mouthfeel, notes of dried herbs, and a caramelly, almost vanilla edge whose sweetness is reminiscent of cognac.

*No added sulfites

Gut Oggau, *Winifred*

Burgenland, Austria

Blaufränkisch, zweigelt

Blueberry | Red cherry | Cinnamon

Stephanie and Eduard Tscheppe-Eselböck make an inspired range of wines in the Burgenland, eastern Austria. This serious rosé is a savory, grown-up, controlled wine. Red- and purple-berry driven, it shows some dark spices and a lovely density (thanks in part to a little tannin). It is wine with a bit of a crunch, as well as a slight coolness that means it can hold its own against a lot of different foods.

*No added sulfites

Domaine Ligas, *Pata Trava Gris*

Pella, Greece

Xinomavro

Blood orange | Pine | Woodland strawberry

Without a shadow of a doubt, the formidable Ligas family are Greece's foremost (commercial) natural grower-makers. Based in northern Greece, in the land of Macedonia and Alexander the Great, the Ligas vineyards are committed to permaculture and to the renaissance of old, indigenous Greek grape varieties. In fact, they've even dedicated a whole series of wines to them, with some of the cuvées produced in tiny amounts—no more than a single barrel. This xinomavro is earthy, full-bodied, and, if you close your eyes and take a sip, you can almost hear the cicadas chirping. A real taste of the Med.

*Low levels of sulfites added

Julien Peyras, *Rose Bohème*

Languedoc, France

Grenache, mourvèdre

Watermelon | Orange Blossom | Raspberry

With Bernard Bellahsen of Fontedicto (see *Horses*, pages 106–107) as your mentor, you simply cannot go wrong, and Julien Peyras is lucky to have learned from one of the pillars of natural wine. What's more, it shows. Julien's wines are alive and mouth-wateringly good. Made from 70-year-old grenache (grown on basalt) and 10-plus-year-old mourvèdre (grown on clay), this wine is what is called a *rosé de saignée*, which means that part of the juice has been bled off the must after a short maceration (in this case, 24 hours), resulting in a deeply colored, flavorsome rosé.

*No added sulfites

Borachio, *Flat Out Rosé*

Adelaide Hills, Australia

Field blend

Blood orange | Watermelon | Herbal

Alicia Basa and Mark Warner left Sidney in 2015 and ended up doing a vintage at Jauma for renowned natural wine producer, James Erskin. The rest is history, as they say, because they never left the Hills. They buy in fruit from organically farmed vineyards and produce 20,000 or so bottles, but they're now on the lookout for vineyards to lease with the aim of ultimately owning some land. As Marks puts it, "Our goal is to grow grapes,

Right: **Anne-Marie Lavaysse and her son Pierre produce entirely sulfite-free wines in the rocky Saint-Jean-de-Minervois, France. (See *Medicinal Vineyard Plants*, pages 52–53, for more on the Lavaysses' story.)**

as simply buying fruit and making the wine gets boring. It's just hard to get a foot in the door."

Flat Out Rosé is a medley of different grapes—red and white. There's cabernet sauvignon, pinot noir, merlot, chardonnay, pinot gris, and savagnin. The result is ridiculously juicy, flavorsome, and easy-drinking. However, the "fruit-salad" approach is coming to an end with the 2019 vintage, as Alicia and Mark have started using only merlot grapes, which were originally used as fodder for the farm's horses. "It was such a waste, so we decided to make something fun and strange out of them instead," says Mark, "somewhere between a light red and a rosé." He may yet add some pinot noir back into the mix, but what is certain, is that "the style and the sentiment behind the wine will remain the same."

*No added sulfites

FULL-BODIED PINKS

Strohmeier, *Trauben, Liebe und Zeit Rosewein*
Weststeiermark, Austria
Blauer wildbacher

Licorice | Bilberry | Savory rose petal

Franz Strohmeier is an extraordinarily daring, talented winemaker. Living in an area known for its schilcher (tart rosé wines, whose malo is usually blocked), which are, in my opinion, often bland and boring, Franz goes against the grain. Copper-colored, his rosé shows a deeply expressive licorice nose, with wild blueberry and savory rose-petal notes on the palate, that also develops secondary, earthy aromas. Fresh, with a crunchy texture, it is surprisingly youthful given its age.

*No added sulfites

Les Vins du Cabanon, *Canta Mañana*
Roussillon, France
Grenache blanc, grenache noir, carignan, mourvèdre, muscat

Rose petal | Strawberry | Poppy

Alain Castex is a stalwart of the natural wine world. The creator of the exceptional Le Casot des Mailloles (see *White Wines*, page 151), Alain sold his vineyards up above Banyuls, keeping only those in Trouillas, including my personal Desert Island wine: *Canta Mañana*. Grown on land set back from the coast, in the foothills of the Pyrenees, this pink is a field blend of red and white grapes, and is one of the most expressive pinks I know. If you are of the opinion that rosés are personality-less and meant for thoughtless drinking, think again. This is a wine with a big presence. Very aromatic and grapey, it is round, full, and more than a little spiky.

*No added sulfites

Domaine de L'Anglore, *Tavel Vintage*
Rhône, France
Grenache, cinsault, carignan, clairette

Tangerine | Cinnamon | Gingerbread

Beekeeper-turned-winemaker Eric Pfifferling is perhaps the reference when it comes to pinks. He makes some of the most exciting rosés around and they are famed for their ability to age. His *Tavel Vintage* is pleasurably drinkable, but with a power and weight that put it in a league of its own. Intense, long-lived, and a little zesty, this is rosé at its most profound.

*Low levels of sulfites added

Opposite: **When Alain Castex sold Le Casot des Mailloles, he retained the vines that make *Canta Mañana*, and now bottles this wine under the label Les Vins du Cabanon.**

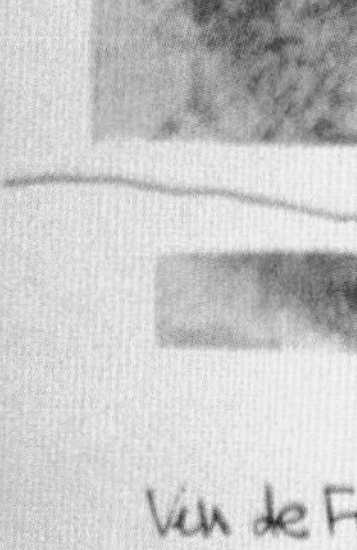

Canta Mañana
L 12 02
Poudre d'Escampette
L 12 03
VENDANGE 2012
Le Casot des Mailloles
Vin de France
de Ghislaine Magnier et Alain Castex
Mis en Bouteille à la Propriété - F.66650
13% vol.
Tél. 04 68 88 59 37
75 cl
- No sulfites -

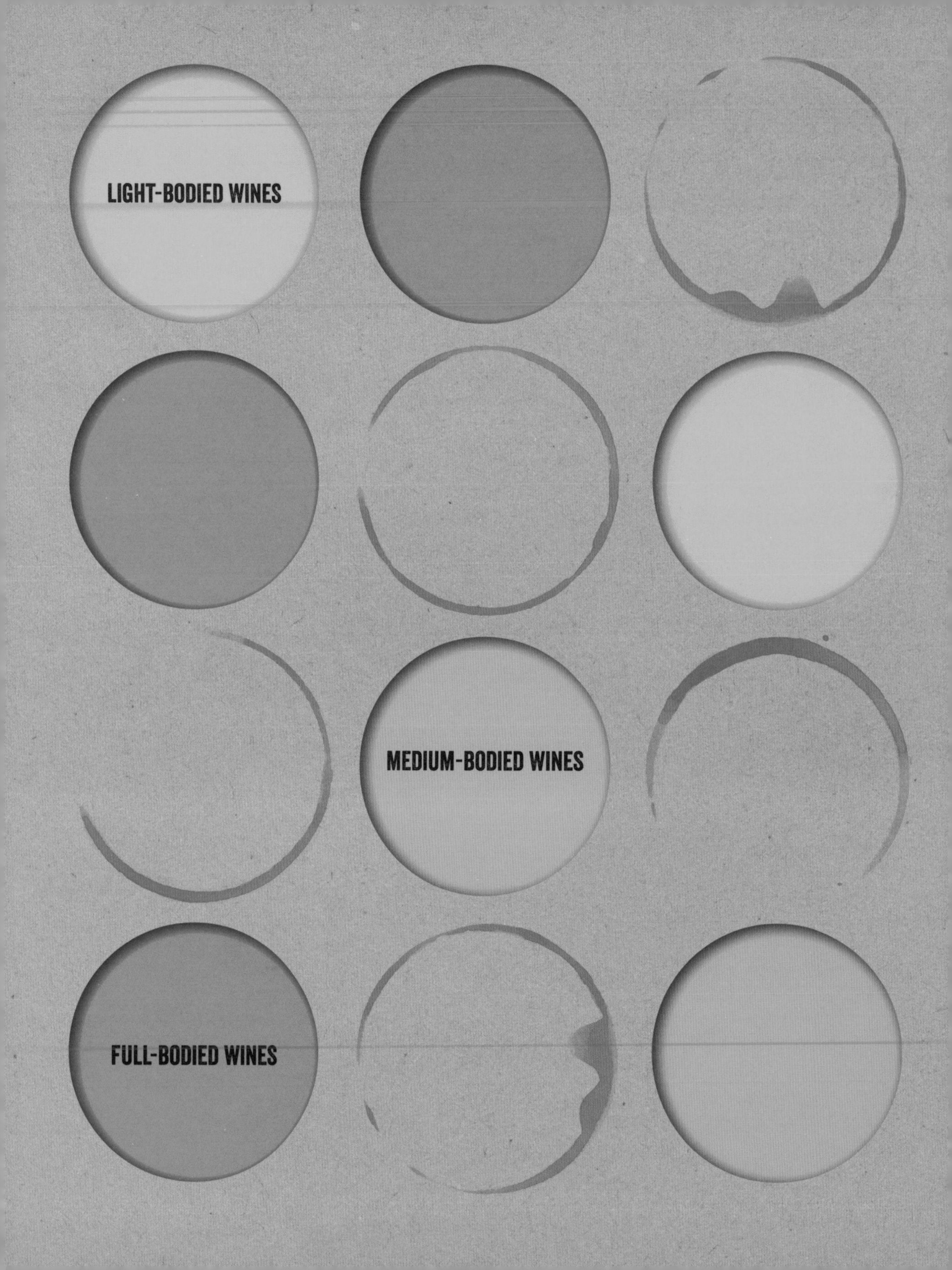
LIGHT-BODIED WINES
MEDIUM-BODIED WINES
FULL-BODIED WINES

Natural reds show aroma profiles that are not as radically different from those of their conventional counterparts, as is the case with white and orange wines. The main reason for this is that still, conventional reds are generally made more naturally than any other wine color or style (with the exception of added yeasts, which still remains the norm). This is because, as with natural reds, the juice is left to stew on the skins (and/or pips and stems), so that the color leaches out. With it come tannins and antioxidants that help protect the wine from oxygen. As a result, conventional growers tend to add fewer sulfites to their red wines than they would to their white wines. Even the legal EU limits for the total amount of sulfites allowed in wine are lower for reds than for whites.

RED WINES

Having said that, there is a distinctive naturalness to the wines, which you will definitely be able to taste. Firstly, you rarely find natural wines caked in aromas of new oak, especially not deliberately. Finding good-quality old barrels can be tricky, however, and, more often than not, producers end up having to buy new barrels that they then season themselves. This means that their first few vintages can show more oak notes and coarser tannins. But, apart from this practical measure, natural producers tend to shy away from oakiness, as they see it as a distraction from the purity of their grapes and *terroir*. In *La Renaissance des Appellation*'s charter of quality, for example, 200% new oak—that is, using two new oak barrels during winemaking, which certain conventional wine producers take pride in doing—is forbidden.

Equally, the berries of natural growers tend to be harvested at full phenolic ripeness, but without leaving them to hang in the vineyard for the sake of building up jammy notes, as can be the fashion. Since there is no recourse to rectifying acid, growers have to make sure their grapes naturally retain enough to obtain a good natural balance.

It is also becoming increasingly popular among natural growers to return to the traditional method of fermenting with whole bunches of grapes (i.e. to ferment everything together rather than de-stemming the grapes first). If the stems are perfectly ripe, this adds extra complexity of texture, freshness, and floral notes that are almost violet-like. Antony Tortul, one of whose wines is featured in this selection and who makes a crazily diverse, fantastic range of wines without any additives, is a grower who uses this method. In fact, he does not use any temperature control, even in the height of summer. Indeed, I loved his simple, matter-of-fact reasoning when he explained to me that his wines, "Come from the South of France, where grapes see 35°C for three months of the year. This means that, by harvest, they have learned to withstand the heat. They are used to it. Some wines go up to 30°C during fermentation, but it does not worry me. The last thing I want is to make wines that taste of the vinification, because I try so hard to express the *terroir* in all my cuvées."

Finally, there's the drinkability and moreishness. This is, for growers, at the core of their natural-wine production, and is particularly obvious if you compare natural and conventional reds. Even if the wines are complex and intense, good natural red wines will always show immense freshness and digestibility, no matter how youthful or mature they are—and this is what makes them so compelling.

Below left: **The talented Antony Tortul, and team, in their cellar outside Béziers.**

Below: **Natural wines are such a reflection of their environment that they inevitably become more complex as the diversity of life forms around the vines expands. Twenty or so years ago Claude Courtois, in the Loire, France, noticed such a difference on one plot that it prompted him to create a brand new cuvée—*Racines*.**

Opposite: **A few of Antony Tortul's multitudinous cuvées.**

la sorga
BRUTAL

FRANCE

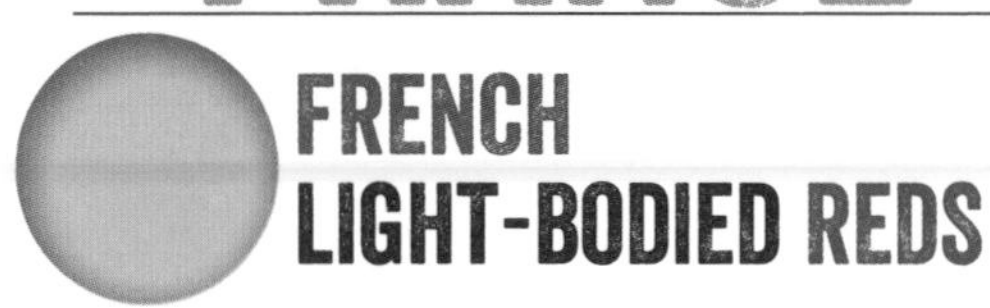

FRENCH LIGHT-BODIED REDS

Domaine Cousin-Leduc, *Le Cousin, Le Grolle*

Loire

Grolleau gris, grolleau noir

Peppery | Poppy | Curry leaf

The now semi-retired Olivier Cousin, a keen sailor and legendary horseman, is perhaps natural wine's most lionized wild child. Having publicly fought the appellation system for many years (see *Who: The Artisans*, pages 100–105) and been uncompromisingly natural with his farming and vinification methods, he has inspired many young growers to opt for this way of life. Probably the wine world's most outspoken advocate of solidarity, Olivier is the ultimate *bon vivant*. With his focus on clean farming and producing easy-drinking wines for consumption in large quantities (as his email signature says: "Save on energy and preserve the environment: uncork less, drink magnums!"), coupled with his commitment to community and cooperation, Olivier has developed a bit of a cult following, with fans from as far away as Japan returning every year to visit or help.

Fragrant and with a lovely soft tannic grain, this easy, round wine is the perfect way to celebrate the arrival of spring. Drink within a day of opening.

*No added sulfites

Guy Breton, *Chiroubles*

Beaujolais, France

Gamay

Violette | Blueberry | Fresh earth

Guy Breton, aka *Petit Max*, is one of the pillars of natural wine in the Beaujolais region. Having worked with Jules Chauvet (see *Who: The Origins of the Movement*, pages 116–17) and Jacques Néauport (see *The Druid of Burgundy*, pages 118–19), he is part of the Gang of Four who inspired so many natural wine growers, not only in their home region but the world over. Juicy and chiseled, this ethereal red shows great perfume, longevity, and precision. Another favorite is his old-vine *Morgon*.

*Low levels of sulfites added

Patrick Corbineau, *Beaulieu*

Chinon, Loire

Cabernet franc

Fragrant cassis | White pepper | Mace

Patrick Corbineau makes gentle wines that are also powerhouses in terms of complexity and their ability to age. Tense and delicious, this chinon, which spent two years in barrel, is a beautiful wine. Patrick's production is tiny and his wines tricky to get hold of, but they are very rewarding if you do manage to. Still incredibly young.

*No added sulfites

Pierre Frick, *Pinot Noir, Rot-Murlé*

Alsace

Pinot noir

Orange peel | Violets | Cumin

Jean-Pierre Frick manages his domaine (of a dozen, mainly chalky, plots) with his wife, Chantal, and son Thomas. A pioneer of biodynamics in Alsace, Jean-Pierre converted his estate in 1981, after having already worked his vines organically since 1970. This 100-year-old pinot noir, from a limestone parcel rich in iron (hence its name, *rot-murlé*, meaning "red-walled"), looks pale and light, yet is incredibly fragrant, detailed, and long on the finish.

*No added sulfites

Left: **This Beaujolais by Julien Sunier is another mid-bodied natural red.**

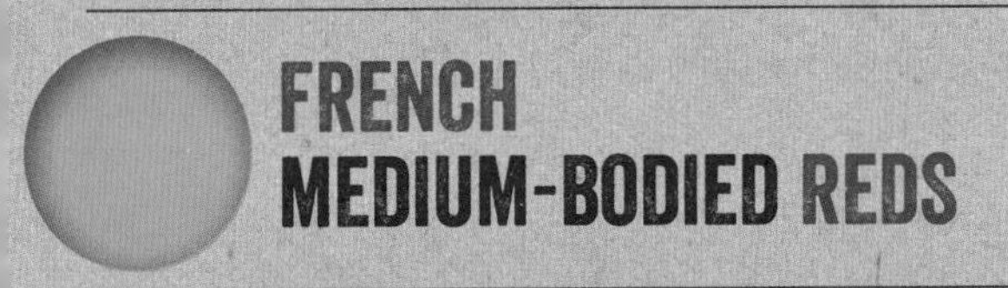

FRENCH MEDIUM-BODIED REDS

François Dhumes, *Minette*

Auvergne

Gamay

Sweetpea | Stony minerality | Mulberry

Inspired by the low-intervention producers of his native Auvergne, including Vincent Tricot (see *White Wines*, page 150), Patrick Bouju, and Stephane Majeune, François Dhumes decided to turn his back on the more industrial teachings of his winemaking school in Burgundy, as well as five years working conventionally in the Rhône. He farms 7 acres (3 hectares) of gamay d'Auvergne and chardonnay on basalt/red clay/limestone soils. The *Minette* has a surprisingly firm, but fine, structure and a bitey minerality, as well as being very scented. A dash of volatile acidity, but nicely integrated.

*No added sulfites

Christian Ducroux, *Exspectatia*

Beaujolais

Gamay

Bilberry | Matcha tea | Alfafa flower

Christian Ducroux is a farmer who is committed to the living above all else. His 12 acres (5 hectares) of vineyard, with its pink granite soil, is a hub of biodiversity: fruit trees alternate with each five rows of vines, hedges surround the parcels, and wild grasses grow throughout. And, as he says on the bottle, "In order to help the plant on its path, [our] mares, Hevan and Malina, help us plow in ways that promote microbial life." *Exspectatia* is soft and juicy to start with, gaining power as it is exposed to oxygen, when it reveals complex, earthy tones and spices, reminiscent of the forest floor. It is a very accomplished wine and anyone in any doubt that gamay is capable of long-lived greatness, or indeed anyone dubious that natural wine can combine refinement, complexity, and purity, then this is a must-taste for you.

Note: As one of the principal birthplaces of the natural-wine rhizome (see *Who: The Origins of the Movement*, page 116), the Beaujolais is a hotbed of great natural growers. A few other growers to look out for as you explore natural wine are Marcel Lapierre, Yvon Metras, Jean Foillard, Guy Breton, Jean-Paul Thévenet, and Joseph Chamonard—the natural wine old guard, if you like—but also a host of newish additions, including Hervé Ravera, Julie Balagny, Philippe Jambon, Karim Vionnet, and Jean-Claude Lapalu, among others.

*No added sulfites

La Maison Romane, *Vosne-Romanée Aux Réas*

Burgundy

Pinot noir

Balsamic | Allspice | Mulberry

Oronce de Beler vinifies no less than 10 different crus in Burgundy from vines he leases from other growers. His wines are edgy, chiseled, and perfumed, and flirt with VA (volatile acidity) in a positive way. Oronce plows by horse, and he and Prosper (his plow horse) lend out their services to other growers. He also started Equivinum, a company that sells horse plows which he designed and developed with the help and know-how of a handful of other Burgundian domaines. Oronce's reds are vinified whole bunch.

*Low levels of sulfites added

Clos Fantine, *Faugères Tradition*

Languedoc

Carignan, cinsault, syrah, grenache

Rosemary | Black cherry | Licorice

Owned and run by a sibling trio, Carole, Corine, and Olivier Andrieu, this 72-acre (29-hectare) farm in southern France (see *Wild Salads*, pages 88–89) is only planted with bush vines, which is a common practice in hot climates. This means that the plants are left to grow like spiders in tufts rather than on wires or posts. The Andrieus' soils are dominated by schist, which is very apparent in another of their wines, *Valcabrières*, a tiny cuvée of pure terret with a biting minerality. *Faugères Tradition*, a carignan-dominated blend, is dark and juicy, with a spicy, savory edge and garrigue notes that remind you of a warm afternoon in the sunny south.

*No added sulfites

Henri Milan, *Cuvée Sans Soufre*

Provence

Grenache, syrah, cinsault

Spicy cherry | Violets | Damson

Located near St Rémy de Provence (made famous by Van Gogh, who spent a year in an asylum there), the family domaine was taken over in 1986 by Henri Milan, who had wanted to be a vigneron since the age of eight when he planted his first vine. After a disastrous first attempt at no-added-SO_2 winemaking, which wreaked financial chaos for him (see *Conclusion*: *A Celebration of Life*, pages 92–95), Henri's butterfly range is today a hugely popular, very-easy-drinking, no-sulfite red; a bestseller in the United Kingdom. Pure and fragrant.

*No added sulfites

Les Cailloux du Paradis, *Racines*

Sologne, Loire

Field blend of dozens of grapes

Forest floor | Red currant | Pepper

Located in the Sologne, where Parisians come to hunt game, Claude Courtois is another natural wine hero (see *Who*: *The Origins of the Movement*, page 114, and *The Artisans*, pages 100–105), and one of the few remaining vignerons in this area of the Loire.

This farm, with its fruit trees, woods, vines, and fields, is a model of agricultural diversity that treats soil with the respect due to any living organism. Dedicated to heritage varieties, the Courtois family even planted syrah after finding references to the variety in a 19th-century text, which told the story of a grower who made what was considered the best red wine of the Loire: a 100% syrah. They initially received the go-ahead from the local authorities, who later recanted, sued the Courtois family, and forced them to grub their vines. The Courtois' *Racines* (see page 180), a multi-blend of grapes, is earthy and complex, showing best after several years as its perfumed, floral side blossoms.

*Low levels of sulfites added

La Grapperie, *Enchanteresse*

Côteaux du Loir, Loire

Pineau d'aunis

Red peppercorn | Nasturtium | Cassis

Having started La Grapperie in 2004 with very little previous winemaking experience, Renaud Guettier is meticulous, both in the vineyard and in the cellar. His 10 acres (4 hectares) are spread over 15 micro plots, each with its own microclimate, and vinified without the use of any sulfites in the winemaking. Instead, Renaud uses the wisdom of time to stabilize his wines. Long *élevages* are the norm here, sometimes for as long as 60 months in barrel. The *Enchanteresse* is precise, elegant, and linear, and a testament to the fact that Renaud's wines are some of the most promising being produced in the Loire today.

*No added sulfites

FRENCH FULL-BODIED REDS

Domaine Fontedicto, *Promise*

Languedoc

Carignan, grenache, syrah

Black olive | Rosemary | Juicy red cherry

Bernard Bellahsen (see *Horses*, pages 106–107), the inspirational, self-taught, incontestable maestro of animal traction, started his agricultural adventure producing fresh grape juice before eventually moving into wine. Today, he also grows ancient varieties of wheat (some 6ft/2m tall), which he and his wife, Cécile, mill and bake into bread to sell at the local market.

Bernard's *Promise* is intense and concentrated, effortlessly challenging the myth that wines without sulfites cannot be aged. A must for the cellar.

*No added sulfites

Jean-Michel Stephan, *Côte Rôtie*

Rhône

Syrah, viognier

Parma violet | Blood orange | Juniper

Fancy tasting a benchmark *Côte Rôtie*? Well, Jean-Michel Stephan's is definitely as good as it gets. *Côte Rôtie* is all he makes, so the full range is worth exploring. Guided by the teachings of Jules Chauvet (see *Who: The Origins of the Movement* , pages 116–17), Jean-Michel has been a natural purist from the start (in 1991): all the work on his steep, organic hillside vineyards is done by hand and his winemaking is low-intervention. He mostly grows old-vine syrah, of which a large proportion is, in fact, sérine, the small-berried, low-yielding local variation of syrah, which is experiencing a resurgence, thanks to the work of a handful of growers, including Jean-Michel.

Although not one of Jean-Michel's famed sérines, this bottle is a classic *Côte Rôtie* blend. It's an expressive, perfumed, ethereal version of syrah with a heady purity, and it is, quite simply, bloody magnificent.

*No added sulfites

Château le Puy, *Emilien*

Côtes de Francs, Bordeaux

Merlot, cabernet sauvignon

Rich plum | Cedar | Cocoa bean

Located on the same rocky plateau as Saint-Emilion and Pomerol, Château le Puy is a rarity in Bordeaux. "One of my grandfathers was too miserly, and the other too much of a visionary, ever to use synthetic chemicals in the vineyard," jokes Jean-Pierre Amoreau, when asked how his Château has managed to farm organically for the last 400 years.

Having shot to fame after their 2003 vintage was featured in the Japanese manga comic *The Drops of God*, the domaine achieved cult status almost overnight. And deservedly so. Le Puy wines are elegant, traditional clarets, which even the most classic of drinkers will enjoy. This 85% merlot is from a ripe year, which brings an added generosity and plushness to what is already an approachable, yet age-worthy, wine.

*Low levels of sulfites added

La Sorga, *En Rouge et Noir*

Languedoc

Grenache noir, grenache blanc

Violet | White pepper | Bilberry

With his laid-back manner and wild, gregarious curls, you'd never guess that Antony Tortul, a trained chemist, is actually über-meticulous by nature (see *Who: The Artisans*, pages 100–105).

This young *négociant*-winemaker in the Languedoc, who started producing in 2008, works with 40 different types of vine, including many heritage varieties, such as aramon, terret bourret, aubun, carignan, cinsault, and mauzac, among others. "I always wanted to produce lots of small batches of pure wines of *terroir* with non-interventionist vinifications," says Antony.

Antony's *En Rouge et Noir* is ethereal, perfumed, and moreish—a superbly pleasurable drinking wine. Definitely a grower to watch.

*No added sulfites

ITALY

ITALIAN LIGHT-BODIED REDS

Cascina Tavijn, *G Punk*

Asti, Piedmont

Grignolino

Red currant | Cherry stone | Juniper

Nadia Verrua's family has grown and made wine on the sandy slopes of Monferrato for over a century. Her 12 acres (5 hectares) are home to hazelnuts and indigenous, old-vine grape varieties, including barbera, ruché, and grignolino. The latter gives us *G Punk*, a wild, bright, tannic (fine) wine with a touch of salinity that also shows a vivid cherry-stone bitterness, which is seemingly characteristic of this ancient Monferrato grape variety. It is thought that "grignolino" is, in fact, derived from the Asti dialect for "many pips" (or *grignole*), which could be the source of some of the bitterness. A deliciously drinkable wine.

*No added sulfites

Left: **Montesecondo's vines in spring, showing the use this vineyard makes of cover crops.**

ITALIAN MEDIUM-BODIED REDS

Cantine Cristiano Guttarolo, *Primitivo Lamie delle Vigne*

Puglia

Primitivo

Bilberry | Balsamic | Lime

People usually think of primitivos (better known in the United States as zinfandel) as huge, gushing wines (which they certainly can be), but Guttarolo's is all about freshness and acidity. Made in a stainless-steel tank, this floral, linear wine is also wonderfully mature and savory.

Located near Gioia del Colle, on Italy's heel, Cristiano also makes an anfora primitivo, which, while tricky to track down, is really worth the effort. It is an amazing wine.

*No added sulfites

Lamoresca, *Rosso*

Sicily

Nero d'avola, frappato, grenache

Mulberry | Violet | Cinnamon

Named after the ancient local "moresca" olives, the Lamoresca estate boasts 1,000 olive trees of its own alongside 10 acres (4 hectares) of vines. Filippo Rizzo pioneered wine production in this area with great results. This blend of nero d'avola (60%), frappato (30%), and grenache (10%) is bursting with bright red fruits.

*Low levels of sulfites added

Selve, *Picotendro*

Aosta Valley

Nebbiolo

Balsamic | Cherry bark | Blackberry

Old-fashioned and rustic, this nebbiolo comes from the smallest, least populous region of Italy, the Aosta Valley in the Alps. Created thousands of years ago during the last Ice Age, this region is the home of mighty peaks, including the Matterhorn, Mont Blanc, and Monte Rosa, and also to Jean Louis Nicco's terraced vineyard. "We have one of the best *terroir* in the world!" he exclaims. Bought in 1948 by Jean Louis' grandfather, who decided to produce natural wine to sell on tap to his fellow villagers, it was then taken over by his mountaineer father, Rinaldo, in 2001, and is now in Jean Louis' hands.

Picotendro (meaning "nebbiolo" in Aostan) is a powerful wine with distinct tannins and a lot of concentration, with marked signs of having spent time in old oak. Long-ageing, the wine has a supple resilience that ages really well (I tasted various older vintages simultaneously).

*No added sulfites

Cascina degli Ulivi, *Nibiô, Terre Bianche*

Piedmont

Dolcetto

Morello cherry | Black olive | Gamey

The late Stefano Bellotti's farm is a model of sustainability—54 acres (22 hectares) of vineyards, 25 acres (10 hectares) of arable land (rotated between wheat and fodder), 2½ acres (1 hectare) of vegetable gardens, 1,000 fruit trees, a herd of cattle, and a collection of other farmyard animals—Stefano farmed organically from the 1970s, converted to biodynamics in 1984, and his team continues his legendary work today. Made from dolcetto grapes (called *nibiô* in the local dialect), which have red stems, this wine is an ancient tradition of the Tassarolo and Gavi regions, with the grape variety having been grown in these parts for more than 1,000 years. With its nicely developed nose, touch of gameyness, and complexing volatility (VA), the tannins are completely integrated into the wine and the result is seamless. Perfectly mature now.

*No added sulfites

Panevino, *Pikadé*

Sardinia

Monica, carignano

Mulberry | Capers | Peppermint

Gianfranco Manca inherited a bakery with its own set of old vines of 30 different grape varieties. And so was born the name *panevino* (meaning "bread wine" in Italian). Thanks to his understanding of breadmaking (and the fermentation this involves), working with grapes was a natural progression. Dense and savory, and completely closed to start with, this quaffable wine opened with dark cherry notes that morphed into more floral, redder ones.

*No added sulfites

Montesecondo, *TÏN*

Tuscany

Sangiovese

Black cherry | Cacao | Orris

Saxophonist Silvio Messana spent years living in NYC, before returning to his family's Tuscan estate on the death of his father (a jazz-musician-turned-wine producer who'd planted the vineyard in the 1970s). At the time, his mother was selling their grapes in bulk, but Silvio dived headfirst into the challenge, bottling his first vintage in 2000.

Nowadays, as Silvio explains, "We think of the winery as a living organism" and the work of winemaking as "a natural transformation." His wines are, indeed, a testament to his philosophy. Named for the Arabic for "clay," *Tïn*—which was made in a 100-gallon (450-liter) Spanish, clay amphora—spent 10 months on its skins and was bottled unfiltered. The result is very floral and elegant.

Did you know? Although Tuscany may not seem synonymous with wilderness, Montesecondo is so remote, you can actually hear wolves howling at night!

*Low levels of sulfites added

ITALIAN FULL-BODIED REDS

Cornelissen, *Rosso del Contadino 9*

Sicily

Nerello mascalese and a dozen or so other local white and red grapes

Wild strawberry | Hyacinth | Pomegranate

Having started his career as a Belgian wine merchant, Frank Cornelissen searched high and low for his perfect *terroir*, ending up on the slopes of the unpredictable Mount Etna, a location that epitomizes his own farming philosophy: "Man can never understand nature's full complexity and interactions." Frank avoids all interventions on the land, choosing to follow nature's clues rather than assert himself over it. He avoids all treatments, "whether chemical, organic, or biodynamic, as these are all a reflection of man's inability to accept nature as she is and will be." The *Rosso del Contadino 9* is simultaneously extremely fun and very serious. Try it to understand.

*No added sulfites

Il Cancelliere, *Nero Né*

Taurasi, Campania

Aglianico

Cassis | Floral | Cranberry vibrancy

Being situated at an altitude of 1,800ft (550m) makes all the difference when you're growing in the Mediterranean's warmer climes. The slight elevation means day and night temperature differences that help lengthen the ripening season, producing generally cooler, less baked wines. Aged for two years in big *botti*, and then for another two years in bottle, the long maturation helps tame aglianico's big, firm structure. The whys-and-wherefores of this process are based on the "peasant art," as vineyard owner Soccorso Romano calls it, of making wine that he learned from his dad.

*Low levels of sulfites added

Case Corini, *Centin*

Piedmont

Nebbiolo

Rose petal | Wild thyme | Morello cherry

The first time I tasted *Centin*, a few years back, I was taken aback by its sublimity. It is, quite honestly, the perfect expression of nebbiolo—a wine with poise, charisma, generosity, and gentleness all at once. A wine, in fact, very much like its creator, Lorenzo.

Lorenzo Corino combines theoretical knowledge (having spent most of his working life researching in the field of agriculture—cereals, viticulture, and winemaking) with practical experience, being the fifth generation of a grower-maker family in Costigliole d'Asti, in Piedmont. He is also an advisor to biodynamic farm La Maliosa (see *Orange Wines*, page 167) in Tuscany. He is an extraordinary man, with a depth of knowledge and a willingness and generosity to share it that is remarkable. After authoring and co-authoring more than 90 technical and scientific publications on viticulture, Lorenzo finally penned his first book—a memoir entitled *Vineyards, Wine, Life: My Natural Thoughts* (published in the spring of 2016) that gathers together all his invaluable experience.

*No added sulfites

Podere Pradarolo, *Velius Asciutto*

Emilia-Romagna

Barbera

Kirsch | Cloves | Balsamic

Podere Pradarolo produces uncompromising wines in the Parma Hills, in Emilia-Romagna. The grapes are fermented without temperature control and with prolonged macerations of 30 days to nine-plus months, regardless of grape color. This barbera spent 90 days on its skins and then 15 months ageing in large oak casks before bottling. A savory, moreish wine.

*No added sulfites

REST OF EUROPE

REST OF EUROPE LIGHT-BODIED REDS

Magula, *Carboniq*
Malokarpatská, Slovakia
Blauer Portugieser
Fresh plum juice | Blueberry | Capsicum

With an incredible diversity of indigenous grape varieties and a plethora of micro-terroirs, countries like Slovakia are very exciting and at the start of their natural-wine journey. Producers with great promise are also beginning to come out of the woodwork, such as Vladimir Magula. "It was easy enough getting my family to embrace organic viticulture, which we did in 2012," says Vlad, "but the transition to natural winemaking was much harder." The generational clash found Vlad having to fight his corner, trying to convince his parents that natural was the way forward. "It is difficult because the fear is very strong—it has even prevented me from going completely sulfite-free, although my experiments are encouraging," he continues. "Low-intervention wine means you have to give up control. You have to have faith in the intrinsic nature of the winemaking process, plus the final wines are different. My mum, for example, does not like cloudy wines. But my dad ended up becoming a huge fan of orange wine. They are now completely convinced of the philosophy, but it takes time in practice."

Heavily inspired by the work of fellow Slovak Zsolt Sütó from Strekov 1075 (see *White Wines*, page 154), Vlad stopped fining and filtering in 2015, and has steadily lowered his sulfites each year. His *Carboniq* was inspired by the Beaujolais method by the same name. It goes through two weeks of whole-bunch carbonic maceration, which Vlad says actually suits the versatile, low-tannic Blauer Portugieser grape, and has 10mg per liter of sulfites added at bottling. The result is a very easy-drinking, upfront wine with crunchy fruit flavors.

*Low levels of sulfites added

REST OF EUROPE MEDIUM-BODIED REDS

Bodega Cauzón, *Cabrónicus*
Granada, Spain
Tempranillo
Blueberry | Pomegranate | Licorice

Made using carbonic maceration, this tempranillo is grown at altitude (Cauzón's vineyards are located at an altitude of 3,500–4,000ft/1,080–1,200m in the Sierra Nevada), where extreme temperatures and slow ripening have a marked effect on the vivacity of color, acidity, alcohol, and tannins of the wine. The vineyard is owned by Ramón Saavedra, a chef from the Michelin-starred restaurant *Big Rock*, on the Costa Brava. He decided to abandon the kitchen and return home to learn how to grow and make wine. Today, Bodega Cauzón is part of a healthy cohort of natural vignerons populating the mountains around Granada. *Cabrónicus* is the lightest and juiciest of its cuvées.

*No added sulfites

Weingut Karl Schnabel, *Blaufränkisch*
Südsteiermark, Austria
Blaufränkisch
Bramble | Floral | Fresh cranberry

"Our doing is based on the principle that we are only guests on our earth," says Karl, "and that our earth needs to be sustained for future generations." This, to the Schnabels, means that such things as land rights or ownership only mean that a person has been given direct responsibility for a part of our shared planet. In other words, explain Karl and Eva, landowners have a duty of care to handle the land for the benefit of the common good, such as through the production of nourishing food, or as a contribution to a healthier planet. This reserved, shy, humble couple are proud farmers, quietly doing really great stuff, simply because it fits with their personal convictions. They build stone piles and create watering holes, for example, to encourage reptilian life in the landscape (including smooth snakes and colubrids). A pure, mineral blaufränkisch, this wine's vibrancy is testament to this way of life.

*No added sulfites

Above: **Spot the vines! Mythopia's living garden in full swing.**

Mythopia, *Primogenitur*

Valais, Switzerland

Pinot noir

Raspberry | Violet | Crunchy red currant

With views of some of the highest summits of the Alps, the steep slopes of Mythopia are a paradisiacal vine garden (see *A Living Garden*, pages 30–31) of wild flowers, fruit trees, leguminous plants, grains, rare birds, green lizards, and more than 60 species of butterfly. Using agricultural practices derived, at least in part, from ancient methods used by the Aztecs, Mythopia is an ecosystem sustained by a rich, symbiotic network of thousands of species of creature. *Primogenitur* is, as Mythopia's owner, Hans-Peter Schmidt, himself describes it, a "coltish, fruity, jaunty" wine with the same upfront enthusiasm as a child who is "growing up in nature, without complexes, or deceptions. A wine that helps you to remember the best of the day." Hear, hear: I couldn't say it better myself.

*No added sulfites

Terroir al Limit, *Les Manyes*

Priorat, Spain

Garnacha

Ripe mulberries | Slatey | Licorice

Dominik Huber makes some of the finest wines in Spain. Having started out without a word of Spanish and no winemaking know-how, it is remarkable to see what he has achieved in a decade. He tills the soil by donkey, harvests earlier than most in Priorat, keeps cuvées vineyard-specific, and whole-bunch ferments ("We don't want to extract—we want to infuse") in large oak barrels. The result is great delicacy and a steeliness that is unusual for modern Priorat.

This garnacha, from 50-year-old vines grown on clay at an altitude of 2,600ft (800m), is pure and tight with dark fruit, gentle tannins, and a very precise, almost chiseled, texture. As with all Dominik's wines, not only is this red surprisingly delicate, an extraordinary feat given its semi-arid location and the pounding Spanish sun, but it also seems to become more and more precise with each new vintage I taste. This is almost certainly the most elegant expression of Priorat you are likely to find.

*Low levels of sulfites added

Costador Terroirs Mediterranis, *La Metamorphika Sumoll Amphorae*

Penedès, Spain

Sumoll

Kirsch | Damson | Rosemary

With vineyards in the foothills of the Pyrenees, some of which climb to an altitude of 3,000ft (900m), Joan Franquet grows 20 varieties of grape (from old vines, some of which are over 100 years old!). These include a host of indigenous Catalan varieties such as dark sumoll, trepat, macabeu, xarel.lo, sumoll blanc, and parellada. This sumoll is well made and considered, having spent nine months in a Spanish tinaja.

*No added sulfites

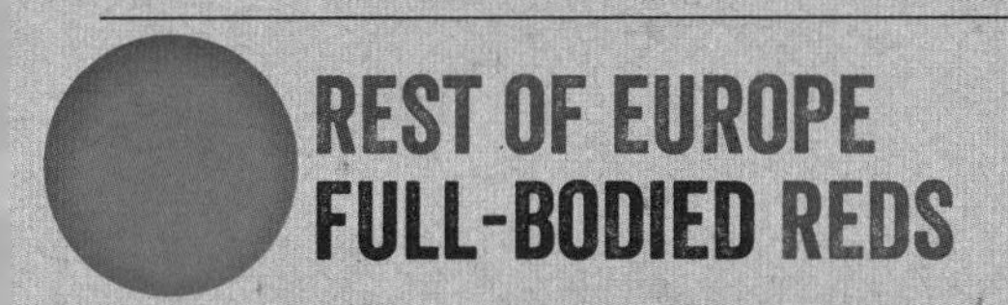

REST OF EUROPE FULL-BODIED REDS

Clot de Les Soleres, *Anfora*

Penedès, Spain

Cabernet sauvignon

Blackcurrant | Wild mint | Lily

Carles Mora Ferrer's beautiful farm building dates back to 1880 and is located just inland from Barcelona in the Penedès. Having made his first additive-free vintage in 2008, Carles' wines have gone from strength to strength. This bottling of cabernet sauvignon spent 13 months in anfora, which, while clearly kissed by the Mediterranean sun, also shows a clarity and purity of aromas brought about by a cooling sea breeze.

*No added sulfites

Nika Bakhia, *Saperavi*

Kakheti, Georgia

Saperavi

Blackberries | Rosemary | Cassis

In 2006, Georgian artist Nika Bakhia, who spends much of his time in Berlin, bought a small vineyard of saperavi and an abandoned wine cellar in Anaga, in Kakheti, which is Georgia's largest winemaking region. His 15 acres (6 hectares) of land includes saperavi, rkatsiteli, and a small collection of tavkveri, khikhvi, and kakhuri mtsvane, which he uses for experimentation. "Winemaking is a creative process," he explains, "like sculpting or painting; it is based on the understanding of the nature of materials and not rejecting or oppressing their original features."

Saperavi skin is thick and its flesh tainted, always yielding intensely colored, almost black, wines—in fact, I once dyed a T-shirt in saperavi juice and it came out lilac! Nika's *Saperavi* is a very intense, concentrated, tannic wine, which was vinified and matured in traditional *qvevri* clay pots that are buried in Nika's cellar—a process that was officially recognized by UNESCO as part of the Intangible Heritage of Humanity in December 2013.

*Low levels of sulfites added

Barranco Oscuro, *1368, Cerro Las Monjas*

Granada, Spain

Cabernet sauvignon, cabernet franc, merlot, grenache

Ripe blackberry | Cinnamon | Oak toast

Named after the altitude of the vineyard (4,500ft/1,368m above sea level), Barranco Oscuro's vines are some of the highest in Europe. Growing in the foothills of Sierra Nevada, in Andalucia, the coolness of the location helps create wines of freshness and great acidity, while the blistering sun of southern Spain gently bakes the berries on the vine. The result is muscular wine with dark-berry fruit that is very Spanish in its opulence and ripeness, while also tighter in texture than most. Although oaky (probably the oakiest wine in this section), it has great layering and depth. Needs food on the side.

*No added sulfites

Purulio, *Tinto*

Granada, Spain

A fruit salad: syrah, cabernet sauvignon, merlot, tempranillo, cabernet franc, pinot noir, petit verdot

Rosemary | Black olive | Mulberry

Torcuato Huertas has worked the land all his life, growing mainly olives and fruit. Back then, his wine was solely for home consumption, but in the early 1980s, with the help of his mentor (and relative) Manuel Valenzuela from Barranco Oscuro (see above for more details), his focus began to change. Today he farms some 7 acres (3 hectares), on which he grows a staggering 21 grape varieties. Being a blend of seven varieties, *Tinto* is not about varietal expression, but is rather a product of place. It owes its ripe fruit flavors to the southern heat and its freshness to its altitude.

*No added sulfites

Dagón Bodegas, *Dagón*

Utiel Requena, Spain

Bobal

Prune | Cherry liqueur | Balsamic

Dagón's wines are directly informed by Miguel's singular farming style, which he has developed over the course of decades. Interventions in the vineyards are minimal, with no farming aids whatsoever (including manures or even Bordeaux Mixture) having been used since 1985. Instead, Miguel stood by the logic that vines should adapt to their environment, which, in this case, is the indigenous Mediterranean flora and fauna that share their space. And adapt they did, with Miguel's grapes believed to be among the healthiest in the world (see *Health: Is Natural Wine Better For You?*, pages 84–87).

Bone-dry and as uncompromising as its guardian, Miguel's bobal was macerated on its skins for several months, pressed, and then aged in oak, where it remained for some 10 years before bottling. While not the sort of wine you're likely to drink a whole bottle of in a single sitting, its intensity is utterly delicious. It's more of a meditative type of wine.

*No added sulfites

Els Jelipins, *Font Rubi*

Penedès, Spain

Sumoll, garnacha

Black cherry | Seville orange | Dried herbs

"My story is pretty simple. I liked drinking wine, so I decided to work in it and, once that happened, all I wanted was to make my own wine. This was also, in part, because most of the wines I came across back then were very extracted and powerful, tiring wines, that you couldn't even eat with, they were so big. So I thought, how great it would be to make a wine I actually wanted to drink. And that's one of the reasons I love sumol," says Glòria Garriga, who started Els Jelipins back in 2003. "At the time everyone was ripping it out, because the authorities had declared it a 'lesser' grape, incapable of making quality wine. In fact, it was completely prohibited for use in DO Penedès. But I loved the wines we made from it and I loved the fact that most of the remaining parcels were ancient vines, many over 100 years old and belonging to elderly people who used it only for their own consumption. So, I loved the social side of it and the idea of preserving this heritage."

Thanks to Glòria's work, and the wines she produced, sumol's reputation blossomed so much that the DO started singing to a different tune, extolling the virtues of this magnificent grape, and today new plantings are going back in. Glòria's 2009 *Font Rubi*, each bottle of which is individually painted with a tiny red heart, is rich and opulent, with an incisive minerality and beautiful balance, as well as a touch of volatile acidity that actually brings even more complexity to a very accomplished vintage.

*Low levels of sulfites added

Below: **Named after the Latin *fervere* (meaning "to boil"), fermentation is such a noisy, lively affair that it does look like the juice is boiling.**

NEW WORLD

NEW WORLD MEDIUM-BODIED REDS

Vincent Wallard, *Quatro Manos*
Mendoza, Argentina
Malbec
Blueberry | Violets | Purple basil

This collaborative project gets its name from the four hands of its owners: Emile Hérédia, natural grower and owner of Domaine Montrieux in the Loire, and Vincent Wallard, a French ex-restaurateur from London. Although tricky practicalities, such as sourcing bottles and corks, meant the project was far from easy to put in place, the results are excitingly individual. A polar opposite to most Argentinian malbec, much of which can be formulaic: overripe, lacking juice, and oaky, this part-whole-bunch, part-de-stemmed—a fermentation technique known as the "sandwich method" because of the different stratas in each vat—and unoaked wine shows extra floral, almost exotic, notes, with a pepperyness and softly tannic structure that make it very moreish.

*Low levels of sulfites added

Cacique Maravilla, *Pipeño*
Bio-Bio, Chile
Pais
Black Mulberry | Tamarind | Smoky

Brought to Chile by Spanish missionaries in the mid-16th century, país (or mission as it is known in California) has for too long been considered a second-rate grape variety in Chile where it was traditionally reserved for bulk wine production. Louis-Antoine Luyt (formerly of Clos Ouvert), a French producer who moved to Chile, was the first person to take país seriously and put this grape variety on the map. Fast-forward a few years and now many growers have realized the potential of this extraordinary grape, often hailing from old vines that are sometimes hundreds of years old. Capable of producing dense, long-lived wines, it also makes highly quaffable, juicy examples like this *Pipeño*.

Cacique Maravilla has been in Manuel Moraga Gutierrez's family for seven generations. Grown on volcanic soil, this particular vineyard was planted in 1776, the same year as the US' Declaration of Independence!

*Low levels of sulfites added

Donkey & Goat, *The Recluse Syrah*
Broken Leg Vineyard, Anderson Valley, California, USA
Syrah
Violet | Cinnamon | Black cherry

Jared and Tracey Brandt make wine in a groovy, urban winery in Berkeley, California. Winemakers, rather than wine growers (they buy in grapes), they started their boutique winery in 2004 and have been going strong ever since. Theirs are Rhône-style wines—more "syrah" than "shiraz"; more peppery and crunchy than jammy and alcoholic. They were certainly influenced by their time spent with renowned French low-intervention winemaker Eric Texier (look out for his *Brézème*, *Vieille Serine* from one of the few limestone outcrops of the northern Rhône). This particular bottle was 45% whole-cluster-fermented, spent 21 months ageing in barrel and another 13 months in bottle, as Jared and Tracey believe the wine needs time to evolve before release. Note: some of their cuvées contain more sulfites than others.

*Low levels of sulfites added

Shobbrook Wines, *Mourvèdre Nouveau*
Adelaide Hills, Australia
Mourvèdre
Fleshy cherry | Pomegranate | Bergamot

The Aussie Tom Shobbrook is like a wild and excitable child, impassioned by everything he comes across—coffee, music, even cured meats, which he has taken to making in his cellar. You get the feeling with Tom that there are no limits: whatever mad fancy goes through his head may well become reality and likely turn out delicious. The result is that his winery feels more like a fun space for experimentation, where someone who is clearly in love with food, flavors, and taste lets his fantasies fly. A sensitive winemaker, Tom is firmly in the camp of the new wave of young *terroirists* who have taken root down

Above: **A *tinaja* sample I tasted at Bichi winery in Baja California, Mexico. Bichi (see below) uses large concrete pots for fermentation. These are known in Spanish as *tinajas*.**

under. This mourvèdre is notable not only for being an early-release wine (hence its name), but also because it is bottled with some residual natural CO_2, which gives it a dash of spritz and keeps it young, fun, and fresh.

*No added sulfites

Bichi, *Santa*

Baja California, Mexico

Rosa del Peru

Pomegranate | Blood orange | Stony

Noel Tellez is leading Mexico's natural *revolución* out of Tecate, in Baja California, where he grows grapes biodynamically, buys in other grapes from old vineyards nearby in order to supplement his production, and then makes soulful, precise, yet thirst-quenching wines. His mother is a true lover of nature and runs the biodynamic side of things.

Made from unirrigated, ungrafted, 100-year-old vines of rosa del Peru (aka moscatel negro) planted at altitude on sand and granite, *Santa* is deceptively pale in color as on the palate it is a tense powerhouse laden with smoky pomegranate and great *terroir* definition.

*Low levels of sulfites added

Old World Winery, *Luminous*

California, USA

Abouriou

Mulberry | Malizia cherry | Rooibos tea

Darek Trowbridge, owner and winemaker of Old World Winery, is a museum curator of sorts, into whose care has been placed the last remaining plot of abouriou vines in California. "I am upholding the planting history and tradition of my family, as well as the curiosity of this heirloom variety," says Darek, who has been working these 80-year-old vines on his folks' estate in the Russian River Valley since 2008. Born into an established Sonoma grape-growing family of Italian ancestry, Darek learned "Old-World-style" winemaking from his grandpa, Lino Martinelli. *Luminous* lives up to its name, thanks to its note of bright cherry. (Also look out for Darek's homemade prickly-pear ice cream, which he sells at the cellar door!)

*No added sulfites

Clos Saron, *Home Vineyard*

Sierra Foothills, California, USA

Pinot noir

Sweet pomegranate | Mulberry | Blonde coffee

Having arrived in the back of beyond with fellow members of a spiritual group, Gideon Beinstock eventually went solo with his own farm and vineyard, Clos Saron, named after his wife. "Saron, my inspiration, has many years of experience in viticulture," says Gideon, "and a magic touch with all living things: dogs, cats, chickens, rabbits, bees, even little human kids." Gideon has farmed using the lunar cycle ever since he noticed that the movement of lees he was caring for in a glass-sided barrel seemed to correlate with particular points of the month. Produced in tiny quantities (just 852 bottles, in fact), this 2010 reminded me of Gideon himself, a man who doesn't speak just for the hell of speaking. The *Home Vineyard* does not gush either—you have to make the first move, but then you're rewarded with lots of hidden personality. A wine with floral prettiness and a great skeleton, it is tight and restrained, and more than a little reserved.

*No added sulfites

Methode Sauvage, *Bates Ranch*

Santa Cruz Mountains, California, USA

Cabernet franc

Crunchy plum | Wisteria | Raspberry leaf

The last time I was in San Francisco, I met up with DC Looney from *Punchdown* (a great natural-wine hangout in Oakland), and walked away with a bottle of this fab wine. I cracked it open the very same day on landing back in the U.K.—so much for natural wine not traveling. Methode Sauvage believe that the idea behind their project is "to find a voice for Californian cabernet franc and chenin blanc from all over the state," and find its voice they certainly have. It sings. A great wine.

*No added sulfites

Montebruno, *Pinot Noir, Eola-Amity Hills*

Oregon, USA

Pinot noir

Wild raspberry | Lily | Stony

Joseph Pedicini grew up in the New York area and, in the early 1990s, set off on a career in the micro-brewing industry, which was in its infancy at the time. By chance, work took him to Oregon, where he drank so much pinot noir that, pretty soon, beer was out and wine was in. "I grew up in a family of folks that emigrated here from Italy—one grandmother from Bari and the other from outside Naples—so were making wine at home, ever since I was a little kid. My grandmother and my father were huge influences on me as far as teaching me fermentation skills, gardening skills, and growing skills for that matter." And, given what Joseph is producing today, I am sure Nonno and Nonna must be very proud. This pinot, growing on a particularly cool spot, thanks to a Pacific Ocean breeze, is fragrant, pure, and awesome.

*Low levels of sulfites added

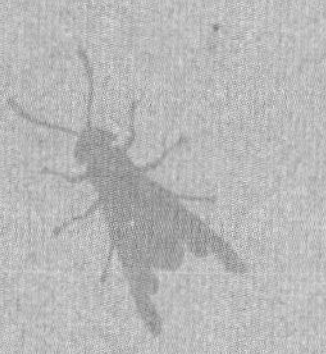

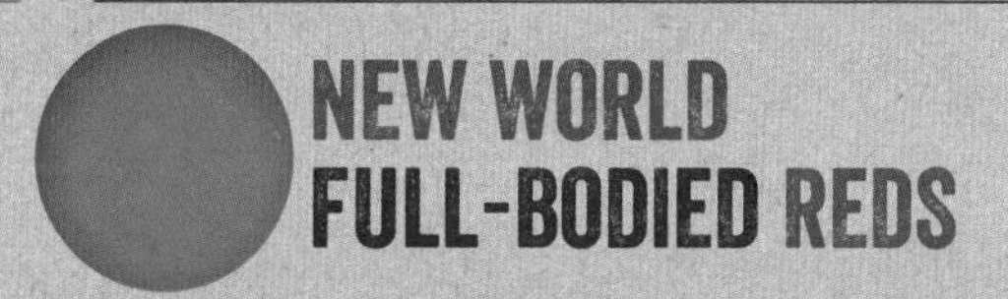

NEW WORLD FULL-BODIED REDS

Castagna, *Genesis*

Beechworth, Australia

Syrah

Blackberry | Violet | Star anise

Located just over 3 miles (5km) outside the historic Victorian town of Beechworth, Castagna's vineyards grow 1,600ft (500m) above sea level, up in the foothills of the Australian Alps. Owned by film director Julian Castagna and his film producer/writer wife, Carolann, the farm was designed along permaculture lines (see *The Vineyard: Natural Farming*, pages 32–37) in order to, in their words, "maximize the use of the land, while minimizing our impact on it." To do this, they turned to David Holmgren (one of the Permaculture Movement's founders), who helped identify key native trees, water-collection points, and the like, which all informed the final layout, and also advised on the building of a straw-bale winery. More than 15 years later, Castagna is today a much-celebrated producer, growing wines that are as alive as their farm and always showing great finesse.

*Low levels of sulfites added

Tony Coturri, *Zinfandel*

Sonoma Valley, USA

Zinfandel

Dark cherry | Crème brulée | Clove

A trailblazer, and often misunderstood, Tony Coturri is Mr. Zinfandel and a grower-cum-maker of very traditional Californian zins. Tony's wines are in no way timid, but, being wonderfully balanced, they express a fascinating California that has sort of fallen off the map. Neither a large, branded wine, nor a trendy, pseudo-European one, Tony's wine is heart-and-soul, proper California. Vastly under-rated, savory, and complex, no United States natural wine bar, store, or listing worth its salt should ignore Tony, his wines, or what he means for America.

*No added sulfites

Bodegas El Viejo Almacén de Sauzal,
Huaso de Sauzal
Maule Valley, Chile
País
Red currant | Black fig | Smoky

Born of very old, ungrafted vines in central Chile, some of which date back to 1650, this país—a grape variety that has seen a recent renaissance—encapsulates traditional Chile. Farmed without irrigation, using techniques that, according to Renán Cancino (the grower-maker), were inherited from the Spanish conquistadors, El Viejo Almacén plows its vineyards by horse and eschews all synthetic fertilizers, agrochemical pesticides, or fossil fuels. Even vinification is traditional—using open-top, oak fermenters and old oak barrels, where the wines remain for a year before bottling. They then wait another year before being released. Bottling and labeling is done by hand.

Sauzal itself was settled back in 1789 by aristocratic families, who owned most of the land surrounding the town. As Renán explains, "My paternal grandmother was a nanny to one of the[se] aristocratic families. She was a single mother with one daughter, Julia, and one son, Bolivar, my father. She later dedicated her time to sewing back home, to raise her children, and was finally able to open her own store (Almacén), where she became a seamstress in 1960. My father followed in her footsteps and worked in the Almacén until the 2010 earthquake that destroyed most of Sauzal."

*No added sulfites

Left and opposite: **Tony Coturri's vineyard and grapes in late summer. (See previous page for more information.)**

LIGHT-BODIED WINES

MEDIUM-BODIED WINES

FULL-BODIED WINES

Sweet wine is created by naturally concentrating grape sugars. This can be done in a variety of ways, including drying grapes on the vine or on racks after harvest; through noble rot (*Botrytis cinerea*), which is a naturally occurring fungus; or by harvesting the berries frozen, a process that creates icewine (or eiswein).

Be it by drying, fungus, or freezing, the result is the same: bags of residual sugar, which means a ready supply of food for yeasts and other creatures. The key issue is then how to stabilize the wine in order to prevent a re-fermentation in the bottle. Sterile-filtering of the juice or the addition of large amounts of sulfites are the easiest and, indeed, the most common ways of doing this, as both methods rid the environment of microorganisms that might cause the wine to continue fermentation. These are the methods of choice of most conventional sweet-wine producers.

OFF-DRY & SWEETS

Natural growers, however, do not have this option. Some use grape spirit to arrest the fermentation and fortify the wine (a process known as *mutage*), since the elevated alcohol stops the microorganisms in their tracks. This is what producers do in Banyuls, Maury, or Port, for example. In terms of the no-added-sulfites route, fortification is certainly the safest and easiest choice, but some natural growers manage to do it without *mutage*, or additives, or heavy manipulation.

Making sweet wines absolutely naturally is a long, slow process, which requires a great deal of patience. Only time can stabilize a sweet wine if no sulfites or sterile-filtration are used. As Jean-François Chêne once told me, "You need grape maturity and about 18 degrees of potential alcohol at harvest—or even 20—as then it is much easier to make *vin liquoreux* without sulfites. If you start with that, then it's just about time. It needs a long *élevage*. Sometimes 24 months, sometimes 36 months, sometimes 5 years or more, depends on the year. Given time, the wine balances itself out and, since the yeasts are in a pretty sugary, alcoholic environment, they struggle and die."

Once bottled, a wine could, in principle, re-ferment, causing tiny bubbles in the wine, which aromatically is not necessarily a problem. In fact, in some instances, it can almost lift the wine. It's just what the drinker makes of it,

Opposite (top left): **La Biancara's recioto grapes drying in the cellar. This traditional process is one way of naturally concentrating the sugars in the grapes.**

Opposite (below): **Vessels like these are traditionally used in Maury and Banyuls, in southern France, for making sweet fortified reds. The wines are left outside in the sun to bake slowly and naturally.**

Above: **The Collectif Anonyme, a modern co-op of friends, makes a collection of red sweet wines on the French/Spanish border, including a couple of Banyuls (fortified) and a *vin naturellement doux* (i.e. without added alcohol) called *Monstrum*.**

and the drinker—used to drinking sterilized, lifeless sweets—could be a little surprised to find a pearly fizz. But, since the majority of natural growers are not driven by the bottom line and will sit on a cuvée for years if necessary until it is ready to face the world, the chance of this is very, very slim. As Chêne explained to me back in 2013, "I still have a few barrels of 2005, which I left because of an imbalance with the sugar, so now I have to wait a very, very long time for everything to settle. I don't plan to add sulfites or to filter, but it means I am going to have to wait a very, very long time."

To be on the safe side, some growers do, however, bottle their sweets under beer cap, in sparkling wine bottles, to be sure that the packaging can withstand any unexpected build-up of pressure should the yeasts decide to kick in again.

Most of the wines listed here were made without any sulfites, a couple have seen light filtration, and another few are fortified. They are all natural. Those made without any sulfites or fortification are truly remarkable—feats of nature that most in the conventional world would tell you are impossible. They were matured for years for stability, and show some of the most profound, complex aromas and textures you will ever have encountered, lingering on the palate for ages afterward. Sip them slowly; they are extremely rare.

NOT THE SAME THING

Don't confuse natural sweet wines with the term *Vin Doux Naturel* (VDN)—which literally translates as "natural sweet wine." This is an official *appellation d'origine protégée* term that applies to any wine in France, including Maury or Banyuls, whose fermentation has been arrested by fortification, but which, confusingly, may not be natural at all since commercial yeasts, sulfites, etc. may be added.

MEDIUM-BODIED OFF-DRY & SWEETS

Esencia Rural, *De Sol a Sol Natural Dulce*
Castilla La Mancha, Spain
Airén, moscatel
Cinnamon | Caramelized nut | Dry citrus peel

Located an hour's drive south of Madrid, Esencia Rural is a 124-acre (50-hectare) farm that grows and makes manchego cheese, herbs for Weleda cosmetics, black garlic, and a variety of grapes, including airén—the most widely planted grape variety in Spain, which usually goes into making Spanish brandy. In this case, however, the raisined airén and moscatel grapes are fermented and macerated for 148 days, then decanted into an amphora prior to bottling (which is done without sulfites or filtering). Each bottle contains 35g residual sugar.

*No added sulfites

Le Clos de la Meslerie, *Vouvray*
Loire, France
Chenin blanc
Ripe pears | Flint | Pollen

American banker-turned-wine-producer Peter Hahn set up in the Loire in 2002. His *Vouvray* is fermented and aged on the lees in barrel for 12 months and spends six months in bottle prior to release. An off-dry, concentrated chenin, it also has a mouth-watering minerality that is incredibly pure. Smoky with notes of wet wool. A big wine.

*Low levels of sulfites added

Les Enfants Sauvages, *Muscat de Rivesaltes*
Roussillon, France
Muscat
Turkish delight | Passionfruit | Grapey

Germans Carolin and Nikolaus Bantlin fell in love with the south of France. They gave up their city jobs and moved their family to Fitou more than a decade ago. Having started producing their sweet *Muscat* to satisfy the demands of their German family, the result has today become so popular that they struggle to keep up with demand. Showing primary, grapey flavors, this is a youthful fortified wine with a touch of exotic fruit.

*Low levels of sulfites added

La Coulée d'Ambrosia, *Douceur Angevine, Le Clos des Ortinières*
Loire, France
Chenin blanc
Honeyed almonds | Dates | Lime

Having inherited his 10-acre (4-hectare) estate in the Loire from his parents, Jean-François Chêne began farming organically as soon as he took over in 2005. His *Douceur Angevine*, made from botrytized grapes, which are picked at more than 20% potential alcohol, was matured in barrel without additives or manipulation for five years. The result is extraordinarily nutty and fruity.

*No added sulfites

Domaine Saurigny, *S*
Côteaux du Layon, Loire, France
Chenin blanc
Honey | Walnut paste | Crème brulée

After studying enology in Bordeaux, and having worked a stint as cellar master in the Pomerol/Saint Emilion/Puisseguin region, Jerome Saurigny settled in the Loire, inspired by Les Griottes and the non-interventionist wines they produced. This thick, nectar-like wine has the most amazing texture, a little like liquid honey or even *Tokaj eszencia*. There are notes of walnut paste and crème brulée, and a passionfruit acidity.

*No added sulfites

FULL-BODIED OFF-DRY & SWEETS

Clot de l'Origine, *Maury*

Roussillon, France

Grenache noir and a little grenache gris, grenache blanc, macabeu, carignan

Prune | Blackberry | Mocha

Set up by Marc Barriot in 2004, this 25-acre (10-hectare) domaine in the south of France has vineyards in five different villages of the Agly Valley, each with its own soil and microclimate: Calc, Maury, Estagel, Montner, and Latour de France. Tannic, primary, upfront, concentrated, and pure, this fortified grenache noir wine tastes a bit like fresh grapes and black cherries. This is in large part down to Marc's minute yields (only 8 hectoliters per hectare) and his use of *mutage sur grain*—the traditional, artisan method of making quality-driven fortifieds. *Eau de vie* (grape spirit) is poured over whole berries, so trapping the grape's primary flavors.

*Low levels of sulfites added

Vinyer de la Ruca

Banyuls, Roussillon, France

Grenache

Cocoa beans | Bergamot | Black mulberry

As grower Manuel di Vecchi Staraz writes on his website: "*Tot es fa a la mà.*" This Catalan vineyard produces only 1,000 bottles a year (all of which are hand-blown), without any electric or petrol-powered machinery whatsoever—"nothing that revolves, slides, engages, or accelerates," as Manuel puts it. This sweet wine, from 50-year-old vines on the steep slopes of Banyuls, on the Spanish border, is aromatically both flowery and intensely dark.

*No added sulfites

Above: **Vinyer de la Ruca's bottles are all mouth-blown, so each one is unique.**

La Biancara, *Recioto della Gambellara*

Veneto, Italy

Garganega

Saffron | Pecan | Allspice

Created by Angiolino Maule, founder and President of the grower association *VinNatur*, this meditative wine is produced by leaving traditional garganega grapes to dry on racks after harvest. An extended fermentation and maturation process follows, and the wine is bottled some three years later. It has great complexity, lusciousness, and an icy freshness from the acidity, and salted caramel and some tannic texture from part-skin maceration. An opulent wine.

*No added sulfites

Left: **Although some do occasionally host open days for tastings by the general public, many of the growers included in *The Natural Wine Cellar* are so small that they do not have the resources to stay open to the public on a permanent basis. Be sure to call or email ahead of any visits to check that they can receive you.**

AmByth Estate, *Passito*

California, USA

Sangiovese, syrah

Morello cherry | Raisins | Mint

Owned by Welshman Phillip Hart and his American wife, Mary Morwood Hart, this 20-acre (8-hectare) vine and olive estate is Paso Robles' first and only biodynamic certified winery. This passito is made by hanging sangiovese and syrah grapes on a clothes line under a tent to dry! Very raisiny in flavor, it also has a refreshingly minty quality.

*No added sulfites

Marenas, *Asoleo*

Montilla, Cordoba, Spain

Moscatel

Apricot jam | Vanilla bean | Raisin

This tiny, bijou vineyard in southern Spain boasts a grand total of 200 vines! The aromatic moscatel is picked in July and dried in the sun for eight days (hence its name: *asoleo*, meaning "sun-dried"), then fermented in a 200-year-old barrel. The result is a deliciously nourishing, honeyed nectar with 400g residual sugar and only 8% alcohol. Yum.

*No added sulfites

Ledogar, *Mourvèdre Vendange Tardive*

Languedoc, France

Mourvèdre

Coriander | Curry leaf | Jaggery

Xavier and Mathieu Ledogar are brothers from wine-producing stock—two sets of great-grandfathers were growers, as was their grandpa, Pierre, and their dad André. Multilayered, with an extremely long finish, their botrytized *Mourvèdre Vendange Tardive* spent 10 years outside, fermenting and maturing. With 80g residual sugar and 17% alcohol—obtained naturally—this is an incredible wine: very sweet with notes of coffee. It should definitely be drunk on its own; it's too good for food.

*No added sulfites

La Cruz de Comal, *Falstaff's Sack*

Texas Hill Country, USA

Blanc du bois

Rooibos tea | Prune | Seville orange marmalade

Located in Texas Hill Country, between Austin and San Antonio, La Cruz de Comal began as a collaboration between celebrated Californian natural winemaker, Tony Coturri (see *Apples & Grapes*, pages 128–29), and his longtime friend-cum-partner-in-crime, Lewis Dickson, a wine enthusiast. Lewis grows hardy US hybrids—black Spanish and blanc du bois (a rarity, as there is, apparently, a total of only 100 acres/40 hectares in existence)—on limestone soils. *Falstaff's Sack* is a singular, fortified wine that really caught my imagination with its quirkiness. A really nice wine.

*No added sulfites

Viña Enebro, *Vino Meditación*

Murcia, Spain

Monastrell

Cassis jam | Prune | Chocolate

This small, family-owned farm grows a variety of produce: almonds, olives, fruit, and 13½ acres (5.5 hectares) of vines. Rainfall here is very low, so Juan Pascual only works with drought-resistant, indigenous grapes such as monastrell (aka mourvèdre), forcallat, and valencí. Located in Bullas, an area of Murcia famed for its big, high-tannin (dry) reds, Juan decided to do things his own way and this savory dessert wine is one of the results. Grapes are harvested in September, as if to make a dry wine, and then air-dried indoors for six weeks, to concentrate the sugars. The whole bunches are then pressed (stems and all), leaving the juice to macerate on skin for a further three months before moving into old French oak, where it stays for a further four years before bottling. The result is unctuous, concentrated, and tannic with a deep ripeness; this is a wine with backbone and balance. The fruit profile is incredibly ripe—jammy, raisined, and dried fig—so there is a distinct sensation of sweetness. And, although it is perhaps technically incorrect to include the wine in this section, I feel it belongs here, alongside wines that you're likely to have after a meal. It is a wine to have on its own, a wine "to meditate on" (hence its name). It is a dish in itself.

*No added sulfites

Jolly-Ferriol, *Or du Temps*

Vin Doux Naturel, Roussillon, France

Muscats à petits grains, muscat d'Alexandrie

Honey-roasted walnut | Lime confit | Salty caramel

When husband-and-wife team Isabelle Jolly and Jean-Luc Chossart took over one of the oldest viticultural properties in the Agly Valley, they came across a hidden cellar containing very old demi-johns and bottles filled with sweet muscats, which were about 50 or 60 years old. Some were spoiled, but others were magnificent. Jean-Luc decided to use some of these to start his own Vin Doux Naturels. He used old barrels, added 2–4 gallons (10–20 liters) of this old nectar to each, and topped them up with his own 2006 and 2007 VDNs. He used two-thirds muscats à petits grains (to bring finesse) and a third muscat d'Alexandrie (for its intense, grapey aromas), and left the barrels untouched (without topping them up) for 12 years. The result is a rare, unique wine—a natural VDN (see *Not the Same Thing*, page 200) that is "outside time" or *or du temps*, hence its name.

Isabelle and Jean-Luc no longer farm the vines on their property, which they sold when Jean-Luc retired, but they kept the stock of their many bottles of sweet wines, so grab one if possible while they are still available.

*No added sulfites

CO-FERMENTS: A WORLD OF RAINBOW DRINKS

Many natural wine producers end up having to work outside the system, whether they like it or not (see *Who: The Artisans*, pages 100–105). The flipside, though, is that they are free of the restraints many of their conventional counterparts are forced to abide by. This gives producers freedom to be playful with their creations, and is one of the main reasons that the natural wine world is able to be so innovative in its approach.

Producers are always pushing the boundary of what great drinks are and what they can be. Some growers make white wine out of red grapes, a practice also known as blanc de noirs, while others blend red and white grapes to make lighter styles of red wines or deeply colored rosés (see Borachio's Flat Out Rosé, page 175). Perhaps even more excitingly, some have ventured into the realm of other fermentables. Grapes are, after all, not the only foodstuff capable of making fine alcoholic drinks.

Apples make cider, pears make perry, and honey makes mead, to give but a few examples. While these are not new drinks, what is new is that many producers in recent years have been rediscovering heritage ways of doing things. This has meant a return to the use of organic fruit and wild fermentation, and a veering away from practices like pasteurization or the excessive use of sulfites. Many of the practices now employed in the craft versions of these drinks mirror the practices used in the natural wine world. In fact, many natural (grape) wine producers now make a variety of other natural (non-grape) wines too.

It was only a matter of time before cross-pollinated drinks made a debut too—grapes fermented and flavored with flowers (such as elderflower or acacia) or other fruit (like raspberries, plums, or elderberries), or even co-fermented from the start with other edible foodstuffs such as apples or pears. Sometimes the results are co-fermentations through and through, sometimes they are more of a maceration, and sometimes a bit of both.

Some of the drinks are fermented to dryness, creating still wines, but many seem to take the form of bubbles—made in a pet-nat style or using a version of the Traditional Method (see pages 137-139) where the base wine (which may have been made of grapes only or might itself be a concoction of various edibles and aromatics), is then re-fermented with the addition of fresh fruit juice—whatever that fruit juice might be. The possibilities are endless.

Côme Isambert, a small négociant wine producer in the Loire, started playing around with other fruit in 2017. "I started using the Ancestral Method to blend different types of fruit together to improve the quality of juice that might otherwise not have been of sufficient quality to make good wine. By using the other fruit, I was able to create something that was aromatically more interesting," explains Côme. "Also, when I started doing this, we had had difficult vintages here in the Loire. Severe frosts had badly affected the vineyards, but the apple and pear trees had been spared, and were literally bent double under the weight of their fruit, so it was a great opportunity to increase my production."

Depending on the ratio of grapes to other fruit, the resulting drinks are often lower in alcohol than their grape-only counterparts, with totals of around 8.0% to 10.5% ABV. With many modern drinkers looking for drinks with a low ABV, it is perhaps no surprise that many of these co-ferments have gone down a storm. "In the USA our Fusion Cider just flew off the shelves," says Niklas Peltzer from Meinklang, an inspiring, 5,000-acre (2,000-hectare) farm in southeastern Austria, which is a model of biodynamic polyculture and living proof that clean, respectful farming can be done at scale. Meinklang makes beer, juices, and cider, as well as wine. "We had very little experience fermenting apples, so we decided to blend apple and grape juice," explains Niklas, "and then we met Fruktstereo and learned from them."

Fruktstereo is a Swedish Fruit Nat producer—a term they coined to describe their pet nat-style fruit wines. "Initially we called them ciders, but the law prevented us from doing it," explains Karl Sjöström, who, along with Mikael Nypelius, started Fruktstereo. Both Karl and Mikael come from a wine background, so have always treated their fruit as wine producers treat grapes, applying the same natural winemaking logic that their natural wine heroes do in the cellar. They even use a wine press. Having started in 2016 with 660 gallons (3,000 liters), their project has grown quickly to a total yearly production of 11,000 gallons (50,000 liters). All their fruit comes from abandoned orchards, friends' gardens, or is waste fruit from farmers who have been unable to sell off their stock. And, again like wine producers, they separate their fruit in terms of quality, using different batches for different cuvées.

As well as co-ferments, there are many excellent producers worldwide who craft "single variety" non-grape wines, such as ciders, perries, meads, and beer, in much the same way as natural wine growers do, including Arrowood Farms Brewery (beer and fruit beer, New York state, USA); Cyril Zangs (cider, Normandy, France); Fable Farm Fermentory (cider and botanical cider, Vermont, USA); Ferme Apicole Desrochers (mead, from their own hives, Québec, Canada); and Gotsa Wines (cider and beer, as well as wine—see page 141—Kvemo Kartli, Georgia).

Côme Isambert, *Tour de Fruit*

Loire, France

Blend of cooking, eating, and cider apples, grolleau gris and grolleau noir grapes

This was Côme's first foray into the world of not-just-grape fermentation, which he did in partnership with Fruktstereo. A blend of 70% cooking, eating, and cider apples and 30% grolleau gris and grolleau noir grapes, Tour de Fruit is surprisingly vinous and incredibly refreshing. Côme's current repertoire of edibles includes carrots, raspberries, strawberries, quince, pears, apples, and, of course, grapes, so there are plenty more bottles and variations to try.

*No added sulfites

Meinklang, *Fusion Cider*

Burgenland, Austria

Topaz apples, Grüner Veltliner grapes

Homegrown Topaz apples are fermented to dryness, then racked to remove the sediments. Grüner Veltliner grape must is then added and the whole lot is bottled while still fermenting to create the pet nat-style bubble in the bottle.

*No added sulfites

Fruktstereo, *Plumenian Rhapsody*

Scania, Sweden

Plums (Victoria or mixed varieties), apples (mixed varieties), Rondo grapes

This is the first co-ferment Fruktstereo ever made. Plums go through carbonic maceration, and are then pressed, and fresh apple juice is added to raise the sugar level. Grape juice from whole-bunch, macerated rondo grapes (harvested from a local farm) is then also added, and once fermentation gets the residual sugar down to 8–10g, the clear juice is racked off and bottled for the Ancestral Method to work its magic.

*No added sulfites

RECOMMENDED WINE GROWERS

Below is a list of growers who are, to the best of my knowledge, organic or biodynamic, and who use low intervention in the cellar. I have also included those with wines listed in Part 3, with page references, for ease of reference. Many of the grower-makers listed are totally natural and use no additives whatsoever in their winemaking; others use sulfites and, in some cases, the totals in particular cuvées may exceed 50mg per liter, so please check with the individual grower-maker to find out more. This is not meant to be an exhaustive list (so apologies if a grower is missing), but a collection of interesting producers for further tasting.

ARGENTINA

Vincent Wallard (page 193)

AUSTRALIA

Bindi Wines
Bobar
Borachio (see page 175)
Castagna (see page 195)
Cobaw Ridge
Commune of Buttons
Delinquente Wine Company
Domaine Lucci
Jasper Hill
Jauma
Lucy Margaux Vineyards
Luke Lambert Wine
Manon
Mill About Vineyard
Ochota Barrels
Patrick Sullivan
Revelation
Shobbrook Wines (see page 193)
Sl Vintners (see page 158)
Smallfry Wines
The Other Right

AUSTRIA

Christian Tschida
Gut Oggau (see page 154, 174)
Hager Matthias
Herrenhof Lamprecht
Johannes Zillinger
Koppitsch Alex & Maria
Matthias Warnung
Meinklang (see page 207)
Schmelzer's Weingut
Strohmeier (see page 176)
Weinbau Michael Wenzel
Weingut Alice & Roland Tauss (see page 155)
Weingut Birgit Braunstein
Weingut Claus Preisinger
Weingut Georgium
Weingut In Glanz (Andreas Tscheppe)
Weingut Judith Beck
Weingut Karl Schnabel (see page 189)
Weingut Maria & Sepp Muster (see page 155)
Weingut MG vom SOL
Weingut Werlitsch (see page 156)

BRAZIL

Dominio Vicari (see page 159)

CANADA

Domaine Bergeville
Domaine du Nival
Négondos (see page 166)
Okanagan Crush Pad
Parsell Vineyard
Pearl Morissette Estate Winery
Pinard et Fille
Rigour & Whimsey
Vignoble Les Pervenches (see page 157)

CHILE

A Los Viñateros Bravos
Bodega Montsecano
Bodegas El Viejo Almacén de Sauzal (see page 196)
Cacique Maravilla (see page 193)
Clos Ouvert
Roberto Henríquez
Rogue Vine
Tinto de Rulo
Vida Cycle
Villalobos Wine
Viña Casalibre
Viña Doña Luisa
Viña Maitia
Viña Tipaume
Wildmakers
Yumbel Estación

CROATIA

Giorgio Clai
Piquentum

CZECH REPUBLIC

Dobrá Vinice
Dva Duby
Milan Nestarec
Stawek (Richard Stávek)
Vinarstvi Jaroslav Osicka
Winery Marada

FINLAND

Noita Winery

FRANCE

Alsace

A&A Durrmann
Anders Frederik Steen
Beck-Hartweg
Bruno Schueller
Catherine Riss
Christian Binner
Christophe Lindenlaub
Domaine Barmès-Buecher
Domaine Brand
Domaine Clé de Sol
Domaine de l'Envol
Domaine Geschickt
Domaine Julien Meyer (see page 148)
Domaine Mann
Domaine Vantin Zusslin
Domaine Zind Humbrecht
Jean Ginglinger
Jean-Marc Dreyer
Laurent Bannwarth
Le Vignoble du Rêveur
Les Vins Pirouettes
Pierre Frick (see page 142)
Sons of Wine
Vins d'Alsace Rietsch (see page 141)
Vins Hausherr

Ardèche

Andréa Calek (see page 149)
Daniel Sage
Domaine du Mazel
Domaine Jérôme Jouret
Domaine les Deux Terres

Gregory Guillaume
La Vrille et le Papillon
Le Raisin et L'Ange
Mas de l'Escarida
Ozil Frangins
Sylvain Bock

Auvergne
Aurelien Lefort
Domaine La Boheme
Domaine No Control
François Dhumes (see page 182)
Jean Maupertuis
Marie & Vincent Tricot (see page 150)
Pierre Beauger
Vignoble de l'Arbre Blanc

Beaujolais
Anthony Thevenet
Château Cambon
Christian Ducroux (see page 183)
Christine & Gilles Paris
Christophe Pacalet
Damien Coquelet
David Large
Domaine Clotaire Michal
Domaine de Botheland
Domaine des Côtes de la Molière
Domaine Jean Foillard
Domaine Jean-Claude Lapalu
Domaine Joseph Chamonard
Domaine Leonis
Domaine Marcel Lapierre
Domaine Michel Guignier
Domaine Vionnet
France Gonzalvez
George Descombes
Guy Breton (see page 182)
Jean-Claude Chanudet
Jean-Paul et Charly Thévenet
Julie Balagny
Julien Sunier
L'Épicurieux
Le Grain de Sénevé (Hervé Ravera)
Lilian & Sophie Bauchet
Philippe Jambon
Yvon Metras

Bordeaux
Charivari Wines
Château de la Vieille Chapelle
Château Guadet
Château La Haie
Château Lamery
Château Le Puy (see page 185)
Château Massereau
Château Meylet
Château Mirebeau
Château Valrose
Clos Puy Arnaud
Closerie Saint Roc
Closeries des Moussis
Domaine de Valmengaux
Domaine du Rousset Peyraguet
Léandre-Chevalier
Les Trois Petiotes
Ormiale
Vignobles Pueyo

Bugey
Domaine du Perron
Domaine Yves Duport

Burgundy
AMI
Catherine and Gilles Vergé (see page 149)
Château de Bel Avenir / P.U.R. (also Rhone)
Château de Béru
Clos du Moulin aux Moines
De Moor
Domaine Alexandre Jouveaux
Domaine Ballorin & F
Domaine C & L Tripoz
Domaine Chandon de Briailles
Domaine de la Cadette
Domaine de la Romanée Conti
Domaine de Pattes Loup
Domaine Derain
Domaine du Prieuré Roch
Domaine Emmanuel Giboulot
Domaine Fanny Sabre
Domaine Guillemot-Michel
Domaine Guillot-Broux
Domaine Philippe Valette
Domaine Pierre André
Domaine Sauveterre
Domaine Sylvain Pataille
Domaine Tawse
Domaine Trapet
Domaine Vignes du Maynes
Domaines des Rouges Queues
François Ecot
Frederic Cossard
Jean-Claude Rateau
Jean-Jacques Morel
La Maison Romane (see page 183)
La Soeur Cadette
La Soufrandiere (BRET BROTHERS)
Les Champs de L'Abbaye
Philippe Pacalet
Pierre Boyat (see page 148)
Recrue des Sens (see page 148)
Sarnin-Berrux
Sextant (see page 166)

Champagne
For the purposes of this selection, only certified organic and biodynamic growers have been included.
Champagne Augustin
Champagne Beaufort
Champagne Benoit Lahaye
Champagne Bonnet Ponson
Champagne Bourgeois-Diaz
Champagne Clandestin
Champagne Delalot
Champagne Emmanuel Brochet
Champagne Fleury
Champagne Franck Pascal
Champagne Jacques Lassaigne
Champagne Jérôme Blin
Champagne L & S Cheurlin
Champagne Laherte Freres
Champagne Larmandier-Bernier
Champagne Leclerc Briant
Champagne Lelarge-Pugeot
Champagne Marguet
Champagne Pascal Doquet
Champagne Piollot Pere et Fils
Champagne Ruppert-Leroy
Champagne Val-Frison
Champagne Vincent Couche
Champagne Vincent Laval
Champagne Vouette & Sorbée
David Léclapart
Durdon Bouval
Francis Boulard
Francoise Bedel

Corsica
Antoine Arena
Clos Marfisi
Clos Signadore
Comte Abbatucci
Domaine Giacometti
Nicolas Mariotti Bindi

Jura
Arnaud Greiner
Domaine de l'Octavin
Domaine de la Borde
Domaine de la Pinte
Domaine de la Touraize
Domaine de la Tournelle
Domaine des Bodines
Domaine des Cavarodes
Domaine Didier Grappe
Domaine Houillon (see page 149)
Domaine Jean-François Ganevat
Domaine Julien Labet
Domaine Philippe Bornard
Domaine Tissot
Granges Paquenesses
Peggy Buronfosse

Languedoc
Catherine Bernard
Château de Gaure
Château La Baronne
Clos des Calades
Clos du Gravillas
Clos Fantine (see page 183)
Domaine Beauthorey
Domaine Benjamin Taillandier
Domaine Binet-Jacquet
Domaine Bories Jefferies
Domaine d'Aupilhac
Domaine de Courbissac
Domaine de Rapatel
Domaine des 2 Anes
Domaine des Amiel
Domaine des Dimanches (Emile Hérédia)
Domaine Fond Cyprès (see page 174)
Domaine Fontedicto (see page 185)
Domaine Frederic Brouca
Domaine Guilhem Barré
Domaine Jean-Baptiste Senat
Domaine L'Escarpolette
Domaine La Marèle
Domaine Ledogar (see page 205)
Domaine Léon Barral (see page 151)
Domaine Les Hautes Terres
Domaine Lous Grèzes (see page 149)

Domaine Ludovic Engelvin
Domaine Mas Lau
Domaine Maxime Magnon
Domaine Monts et Merveilles
Domaine Pechigo
Domaine Sainte Croix
Domaine Thierry Navarre
Domaine Thomas Rouanet
Domaine Thuronis
Es d'Aqui
Julien Peyras (see page 175)
L'Etoile du Matin
La Fontude
La Grain Sauvage
La Grange d'Aïn
La Sorga (see page 185)
La Villa Sepia
Le Clos des Jarres
Le Pelut
Le Petit Domaine
Le Petit Domaine de Gimios (see page 150)
Le Quai à Raisins
Le Temps des cerises
Les Cigales dans la Fourmilière
Les Clos Perdus
Les Herbes Folles
Les Sabots d'Hélène
Les Vignes d'Olivier
Les Vignes du Domaine du Temps
Mas Angel
Mas Coutelou
Mas D'Alezon (Domaine de Clovallon)
Mas des Agrunelles
Mas des Caprices
Mas Lasta
Mas Nicot (see page 174)
Mas Troqué
Mas Zenitude (see page 174)
Mouressipe
Mylène Bru
Opi d'Aqui
Rémi Poujol
Vignoble du Loup Blanc
WA SUD
Zélige Caravent

Loire

A la Vôtre!
Benoit Courault
Cave Sylvain Martinez
Chateau du Perron / Le Grand Cléré
Chateau Tour Grise
Clos de l'Elu
Clos du Tue-Boeuf (Thierry Puzelat)
Côme Isambert/Clos Cristal Closed (see page 207)
Cyril Le Moing
Damien Bureau
Damien Laureau
Didier Chaffardon
Domaine Alexandre Bain (see page 151)
Domaine Bobinet
Domaine Breton (see page 142)
Domaine Chahut et Prodiges
Domaine Cousin-Leduc (see page 182)
Domaine de Bel-Air (Joel Courtault)
Domaine de Belle Vue
Domaine de l'Ecu
Domaine de la Coulée de Serrant
Domaine de la Garrelière
Domaine de la Sénéchalière
Domaine de Montcy
Domaine de Veilloux
Domaine des Maisons Brûlées
Domaine du Closel
Domaine du Collier
Domaine du Mortier
Domaine du Moulin (Hervé Villemade)
Domaine du Raisin à Plume
Domaine Etienne & Sébastien Riffault (see page 151)
Domaine Frantz Saumon
Domaine Gérard Marula
Domaine Grosbois
Domaine Guiberteau
Domaine la Paonnerie
Domaine la Taupe
Domaine Le Batossay (Baptiste Cousin)
Domaine Le Briseau
Domaine Les Capriades (see page 143)
Domaine Les Chesnaies (Béatrice & Pascal Lambert)
Domaine les Roches
Domaine Lise et Bertrand Jousset
Domaine Mathieu Coste
Domaine Nicolas Reau
Domaine Patrick Baudoin
Domaine Pierre Borel
Domaine René Mosse
Domaine Saint Nicolas
Domaine Saurigny (see page 202)
Domaines Landron
François Saint-Lô
Herbel
Jean-Christope Garnier
Jérôme Lambert
Julien Courtois (see page 149)
Julien Pineau
La Coulée d'Ambrosia (see page 202)
La Ferme de la Sansonnière
La Folie LuCé
La Grange Tiphaine (see page 140)
La Grapperie (see page 184)
La Lunotte
La Porte Saint Jean
Laurent Herlin
Laurent Saillard
Le Clos de la Meslerie (see page 202)
Le Picatier
Le Sot de L'Ange
Les Cailloux du Paradis (see page 184)
Les Tètes et Domaine des Hauts Baigneux (see page 140)
Les Vignes de Babass (see page 142)
Les Vignes de l'Angevin (see page 143)
Les Vins Contés (Olivier Lemasson)
Manoir de la Tête Rouge
Muriel & Xavier Caillard
Noëlla Morantin
Patrick Corbineau (see page 182)
Philippe Delmée & Aurélien Martin
Pithon-Paillé
Reynald Héaulé
Richard Leroy
Sylvie Augereau
Thomas Boutin
Toby Bainbridge
Vine Revival

Provence

Château Sainte Anne
Domaine de Trévallon
Domaine Hauvette
Domaine Les Terres Promises
Domaine Les Tuiles Bleues
Domaine Milan (see page 184)

Rhône

Clos de Trias
Dard & Ribo
Domaine Arsac
Domaine Charvin
Domaine Clusel-Roch
Domaine de la Grande Colline (Hiratoke OOKA)
Domaine de la Roche Buissière
Domaine de Villeneuve
Domaine des Miquettes
Domaine du Coulet
Domaine Gourt de Mautens
Domaine Gramenon
Domaine Jean-Michel Stephan (see page 185)
Domaine L'Anglore (see page 176)
Domaine La Ferme de Saint Martin
Domaine Lattard
Domaine les 4 Vents
Domaine les Bruyères
Domaine Marcel Richaud
Domaine Matthieu Dumarcher
Domaine Montirius
Domaine Otheguy
Domaine Romaneaux-Destezet
Domaine Rouge-Bleu
Domaine Viret
Domaine Wilfried
Eric Texier
Francois Dumas
La Ferme des Sept Lunes
La Gramière
Le Clos de Caveau
Le Clos des Cimes
Le Clos des Mourres
Le Mas de Casalas
Le Vin de Blaise
Les Champs Libres

Roussillon

Bruno Duchêne
Clos du Rouge Gorge
Clos Massotte
Clot de l'Origine (see page 203)
Clot de l'Oum
Collectif Anonyme
Domaine Carterole
Domaine Danjou-Banessy
Domaine de L'Ausseil
Domaine de l'Horizon
Domaine de l'Encantade
Domaine des Enfants
Domaine des Mathouans
Domaine des Sarradels

Domaine du Matin Calme
Domaine du Possible
Domaine du Traginer
Domaine Gauby
Domaine Gilles Troullier
Domaine Jean-Philippe Padié
Domaine Jolly Ferriol (see page 205)
Domaine Le Bout du Monde
Domaine Le Scarabée
Domaine Léonine
Domaine Les Arabesques
Domaine Les Enfants Sauvages (see page 202)
Domaine les Foulards Rouges
Domaine Potron Minet
Domaine Tribouley
Domaine Vinci
Domaine Yoyo
La Bancale
La Cave des Nomades
La Petite Baigneuse
Le Casot des Mailloles (see page 151)
Le Soula (see page 166)
Le Temps Retrouvé
Les Vins du Cabanon (see page 176)
Mamaruta
Matassa (see page 149)
Riberach
Rié & Hirofumi Shoji
Vignoble Reveille
Vinyer de la Ruca (see page 203)

Southwest
Barouillet
Château Lafitte
Château Lassolle
Château Lestignac
Château Tour Blanc
Domaine Causse Marines
Domaine Coquelicot
Domaine Cosse Maisonneuve
Domaine du Pech
Domaine Guirardel
Domaine l'Originel (Simon Busser)
Domaine Plageoles
Elian Da Ros
Ferme Bois Moisset
L'Ostal (Louis Pérot)
Laurent Cazottes
Mas del Périé
Nicolas Carmarans
Patrick Rols

Savoie
Domaine Belluard
Domaine Prieuré Saint Christophe
Jean-Yves Péron

GEORGIA

Akhmeta Wine House
Alexander's Wine Cellar
Anapea Village
Archil Guniava Wine Cellar
Artana Wines
Baghdati Estates
Baia's Wine
Chateau Khashmi
Chona's Marani
Chveni Gvino (Our Wine)
Dimis Ferdobi
Doremi Wine
Ethno
Gaioz Sopromadze Winery
Gotsa Wines (see page 141
Iago Bitarishvili
Iases Marani
Kakha Berishvili
Khvtisia Wines
Lapati Wines
Lomtadze's Marani
Makaridze Winery
Naotari Wines
Natenadze's Wine Cellar
Nika Bakhia (see page 191)
Nikalas Marani
Nikoloz Antadze
ODA
Pataridze's Rachuli
Pheasant's Tears (see page 168)
Ramaz Nikoladze
Ruispiri Marani
Samtavisi Marani
Simon Chkheidze Wine Cellar
Sopromadze Marani
Tanini
Tsikhelishvili Winery
Zhuka-Sano Wine Cellar
Zurab Kviriashvili Vineyards

GERMANY

Andi Weigand
Collective Z
Das Hirschhorner Weinkontor (Frank John) (see page 141)
Enderle & Moll
Ökologisches Weingut Schmitt (see page 167)
Rudolf & Rita Trossen (see page 156)
Stefan Vetter (see page 154)
Weingut Benzinger
Weingut Brand
Weingut Thomas Harteneck
2Naturkinder (see page 155)

GREECE

Afianes Wines
Domaine de Kalathas
Domaine Ligas (see page 175)
Domaine Tatsis
Georgas Family (see page 156)
Kamara Estate
Kontozisis Organic Vineyards
Sant'Or
Sous le Végétal
Vaimaki Family

ITALY

Abruzzo
Caprera
De Fermo
Emidio Pepe
Lammidia (see page 153)
Marina Palusci
Podere San Biagio
Rabasco
Stefania Pepe
Tenuta Terraviva

Aosta Valley
Selve (see page 187)

Calabria
Cataldo Calabretta
Nasciri

Campania
Cantina del Barone
Cantina Giardino (see page 169)
Cantine dell'Angelo
Casebianche
I Cacciagalli
Il Cancelliere (see page 188)
Il Don Chisciotte (Pierluigi Zampaglione)
Podere Veneri Vecchio

Emilia Romagna
Al di là del Fiume
Cà de Noci
Camillo Donati (see page 143)
Casè... naturally wine
Cinque Campi (see page 143)
Denavolo (see page 167)
Il Farneto
La Stoppa
Maria Bortolotti
Podere Cipolla (Denny Bini)
Podere Pradarolo (see page 188)
Quarticello (see page 140)
Storchi
Tenuta Biodinamica Mara
Terre di Macerato
Vigneto San Vito (Orsi) (see page 153)
Vittorio Graziano

Friuli Venezia Giulia
Damijan Podversic
Dario PrinĐiĐ
Denis Montanar
Franco Terpin (see page 174)
Josko Gravner
La Castellada
Paraschos
Radikon (see page 169)
Vignai da Duline
Villa Job
Vodopivec

Lazio
Abbia Nova
Cantina Ribelà
Cantine Riccardi Reale
Corvagialla
Costa Graia
D.S. bio
Le Coste (see page 153)
Maria Ernesta Berucci
Palazzo Tronconi
Piana dei Castelli
Podere Orto

Liguria
Azienda Agricola Il Torchio
Stefano Legnani
Tenuta Selvadolce

Lombardy
1701 Franciacorta
Az. Agricola Andi Fausto
Azienda Agricola Divella
Azienda Agricola Sorgente Oreste
Barbacàn
Bel Sit
Cà del Vent
Casa Caterina (see page 143)
Fattoria Mondo Antico

Podere Il Santo
Tenuta Belvedere
Vigne del Pellagroso

Marche

Azienda Agricola Maria Pia Castelli
La Marca di San Michele

Molise

Agricolavinica

Piedmont

Alberto Oggero
Azienda Agricola Curto Marco di Curto Nadia
Bera Vittorio e Figli
Carussin (di Bruna Ferro)
Casa Wallace
Cascina degli Ulivi (see page 152, 187)
Cascina Fornace
Cascina Luli
Cascina Roera
Cascina Tavijn (see page 186)
Cascina Zerbetta
Case Corini (see page 188)
Eugenio Bocchino
Ferdinando Principiano
Forti del Vento
La Morella
Lapo Berti Vino
Olek Bondonio
Poderi Cellario
Roagna
Rocco di Carpeneto
San Fereolo
Tenuta Foresto
Terre di Maté
Valfaccenda
Valli Unite (see page 152)

Puglia

Azienda Agricola Francesco Marra
Cantina Pantun
Cantine Cristiano Guttarolo (see page 186)
Fatalone Organic Wines
Natalino del Prete
Progetto Calcarius
Valentina Passalacqua

Sardinia

Meigamma
Panevino (see page 187)
Raica
Tenute Dettori

Sicily

Abbazia San Giorgio
Agricola Marino
Agricola Virà
Aldo Viola
Arianna Occhipinti
Azienda Agricola Francesco Guccione (see page 153)
COS
Dos Tierras Badalucco
Elios
Frank Cornelissen (see pages 168, 188)
Lamoresca (see page 186)
Marabino
Marco de Bartoli
Nino Barraco (see page 152)
Serragghia (see page 169)
Valdibella
Vini Campisi
Vino di Anna
Viteadovest

Trentino Alto Adige

Cantina Furlani
Foradori (see page 167)
GRAWÜ

Tuscany

Ampeleia
Az. Agr. Macea
Azienda Agricola Casale
Azienda Agricola San Bartolomeo
Campinuovi
Casa Raia
Casa Sequerciani
Colombaia (see page 167)
Cosimo Maria Masini
Do.t.e.
Fattoria La Maliosa (see page 167)
Fonterenza
Fuorimondo
I Mandorli
Il Paradiso di Manfredi
La Cerreta
La Ginestra
Macea
Massa Vecchia
Montesecondo (see page 187)
Ottomani
Pacina
Paolo e Lorenzo Marchionni
Pian del Pino
Podere Concori
Podere della Bruciata
Podere Gualandi
Ranchelle
Santa10
Stefano Amerighi
Stella di Campalto
Tenuta di Valgiano
Tunia

Umbria

Ajola
Cantina Collecapretta
Cantina Margò
Fattoria Mani di Luna
Paolo Bea
Raína
Tiberi
Vini Conestabile

Veneto

Azienda Agricola Calalta
Ca' dei Zago
Casa Belfi
Casa Coste Piane
Corte Sant'Alda
Costadilà (see page 142)
Dalle Ore
Daniele Piccinin (see page 152)
Daniele Portinari
Davide Spillare
Del Rèbene
Gianfranco Masiero
Il Cavallino
Il Monte Caro
Il Roccolo di Monticelli
Indomiti
La Biancara (see page 203)
Meggiolaro Vini
Menti Giovanni
Monte dall'Ora
Monteforche
Nevio Scala
Sieman
Società Agricola Il Sasso
Tenuta l'Armonia
Valentina Cubi
Vigna San Lorenzo (Col Tamarie)
Villa Calicantus

JAPAN

Atsushi Suziki
Beau Paysage
Grape Republic
Domaine Oyamada
Domaine Takahiko

MEXICO

Bichi Winery (see page 194)
Bodega dos Buhos

NETHERLANDS

Domein Aldenborgh

NEW ZEALAND

Alex Craighead Wines
Cambridge Road
Hermit Ram
Pyramid Valley
Sato Wines (see page 158)
Seresin Estate

POLAND

Dom Bliskowice

PORTUGAL

Antonio Madeira
Aphros Wine
Humus (Quinta do Paço)
João Tavares De Pina
Quinta da Palmirinha
Vale Da Capucha
Vitor Claro

ROMANIA

Weingut Edgar Brutler

RUSSIAN FEDERATION

UPPA Winery (see page 156)

SERBIA

Francuska Vinarija (see page 154)
Oszkár Maurer

SLOVAKIA

Kasnyik Family Winery
Magula Family Winery (see page 189)
Mátyás Family Estate
Organic (Strekov)
Slobodné Vinárstvo
Strekov 1075 (see page 154)

SLOVENIA

Batič
Biodinamična kmetija Urbajs
Domačija Butul
Klabjan
Klinec
Kmetija Štekar
Mlečnik (see page 167)
Movia (see page 142)
Nando
Vina Čotar (see page 169)
Vino Suman

SOUTH AFRICA

Intellego Wines
Môrelig Vineyards (Wightman & Son)
Mother Rock Wines
Reyneke
Ryan Mostert
Testalonga (see page 169)

SPAIN

Alba Viticultores
Alvar de Dios Hernandez
Alvaro Gonzalez Marcos
Barranco Oscuro (see page 191)
Bodega Clandestina
Bodega F. Schatz
Bodega Frontio
Bodegas Almorquí
Bodegas Cauzón (see page 189)
Bodegas Cueva
Bodegas Gratias. Familia y Viñedos
Bodegas Moraza
Can Sumoi
Carlania Celler
Casa Pardet
Celler de les Aus
Celler Escoda-Sanahuja (see page 166)
Celler Lopez-Schlecht
Celler Succés Vinicola
Clos Lentiscus (see page 141)
Clos Mogador
Clot de les Soleres (see page 191)
Comando G
Constantina Sotelo
Costador Terroirs Mediterranis (see page 190)
Dagón Bodegas (see page 192)
Daniel Ramos
Demencia Wine
Dominio del Urogallo
El Celler de les Aus
Els Jelipins (see page 192)
Els Vinyerons Vins Naturals
Envínate
Esencia Rural (see page 202)
Finca Parera
Frisach
La Furtiva
La Microbodega del Alumbro
La Perdida
Marenas (see page 204)
Mas del Serral
Mas Estela
Mendall (see page 155)
MicroBio Wines
Microbodega Rodriguez Moran
Muchada (Lèclapart)
Naranjuez
Olivier Rivière
Partida creus
Purulio (see page 191)
Recaredo & Celler Credo
Ruben Parera (Celler Finca Parera)
Sedella Vinos
Sexto Elemento
Sistema Vinari
Terroir Al Limit (see page 190)
Uva de Vida
Viña Enebro (see page 205)
Vinos Ambiz
Vinos Patio
Vins Nus
Vinyes de la Tortuga
Vinyes Singulars

SWEDEN

Fruktstereo (see page 207)

SWITZERLAND

Albert Mathier & Fils
La Maison du Moulin
Mythopia (see page 190)
Weinbau Markus Ruch
Winzerkeller Strasser

TURKEY

Gelveri (see page 168)

UNITED KINGDOM

Ancre Hill Estates
Charlie Herring
Davenport Vineyards
Terlingham Vineyard
Tillingham

UNITED STATES

A Tribute to Grace
AmByth Estate (see page 159, 204)
Amplify Wines
Arnot-Roberts
Beckham Estate Vineyard
Bloomer Creek Vineyard (see page 158)
Broc Cellars
Caleb Leisure Wines (see page 161)
Clos Saron (see page 194)
Coquelicot Estate Vineyards
Côte des Cailloux
Coturri Winery (see page 161, 195)
Day Wines
Deux Punx
Dirty and Rowdy
Domaine de La Cote
Donkey & Goat (see page 193)
Edmunds St. John
Eyrie Vineyards
Florèz Wines
Hardesty Cellars (see page 157)
Hatton Daniels
Hiyu Wine Farm (see page 159)
J. Brix Wines
Kelley Fox Wines
La Clarine Farm (see page 158)
La Cruz de Comal Wines (see page 205)
La Garagista (see page 140)
Les Lunes Wine
Lo-Fi Wines
Madson Wines
Maître de Chai
Margins Wines
Martha Stoumen
Methode Sauvage (see page 195)
Montebruno Wine (see page 195)
Old World Winery (see page 194)
Populis (see page 159)
Powicana Farm
Purity Wine
Raj Parr Wines
Roark Wine Company
Ruth Lewandowski Wines
Salinia Wine Company
Sans Wine Co.
Scholium Project (see page 161)
Seabold Cellars
Sky Vineyards
Solminer
Sonoma Mountain Winery
Statera Cellars
Stirm Wine Company (see page 157)
Subject to Change
Swick Wines
The End of Nowhere
Two Shepherds
Unturned Stone Productions
Vinca Minor
Zafa Wines

GLOSSARY

Acetic acid bacteria (AAB)
Bacteria that cause the oxidation of ethanol to acetic acid during fermentation. They are responsible for the creation of vinegar.

Agronomist
An agricultural expert in soil management and crop production.

Alcoholic fermentation
The process by which yeast converts sugar into alcohol and carbon dioxide.

Appellation
A protected geographic area designating the provenance of a wine, be it AOC/AOP (*Appellation d'Origine Contrôlée/Protégée*) in France. Sometimes used in this book as a generic term, which is applied, for example, to Italy and its equivalent DOC (*Denominacion de Origine Controllata*).

Biodynamic farming
A type of very traditional, holistic farming developed by Rudolf Steiner in the 1920s.

Bordeaux Mixture
Copper sulfate, lime and water mixture used as a fungicide.

***Botte* (plural *botti*)**
Italian term meaning a big wine barrel or wooden cask.

Brettanomyces
A yeast strain. When this strain is present in large numbers it can dominate the wine, at which point, aromatically speaking, it becomes a problem, creating overpowering aromas of farmyard or salami.

CFU (colony-forming units)
A unit of measurement used in microbiology to estimate the size of viable bacterial or fungal population.

Chaptalization
Adding sugar to grape juice to artificially produce more alcohol.

Cryoextraction
Process of freezing grapes before pressing. The frozen water contained in the berry is left behind during pressing, thus concentrating the sugar content.

Cuvée
A generic French word used to describe any "batch" of wine, be it a blend or a single bottling.

Disgorgement
The removal of sediments in the final stages of some sparkling wine production.

Enologist
A winemaker

Élevage
French word for the care of wine up until bottling.

Fining
Quickens the precipitation of tiny particles (tannins, proteins, etc.), which are in suspension in the wine, using a variety of agents, including egg white, milk, fish derivative, clay, etc.

Flor
A film of yeasts that can develop on the surface of maturing wine, which is essential in the production of sherry (Spain) and *vin jaune* (Jura), for example.

Foudre
French word for a large oak barrel.

Green Revolution
An agricultural revolution that took place in the mid-20th century, which radically increased total crop production worldwide through technological developments and the use of high-yielding varieties, pesticides, and synthetic fertilizers.

Hectare
10,000 square meters (equivalent to just under 2.5 acres).

Hectoliter
A metric unit of capacity equal to 100 liters.

Indigenous yeast
Also called ambient, this is the yeast population that is naturally present in the vineyard and the winery.

Lactic acid bacteria (LAB)
Bacteria responsible for malolactic fermentation in wine, during which harsh malic acid is transformed into softer lactic acid.

Lees
Sediments made of dead yeast cells and other fermentation residues that collect at the bottom of vats/barrels/bottles.

Maceration
Soaking or steeping grapes in their juice.

Malolactic fermentation
Also known as malo or mlf, malic acid (naturally contained in grape juice) is converted into lactic acid during the winemaking process, sometimes before, but mostly during or after, the alcoholic fermentation.

Mega Purple
Grape concentrate that is used in winemaking to add color and sweetness to a wine.

Mousiness
An off-flavor reminiscent of peanut butter or milk that has gone off.

Must
Freshly pressed grape juice

Mutage
Also called fortification, this is the act of adding spirit to grape juice to stop the fermentation process in its tracks in order to retain natural sugars (used in making port, for example).

Mutage sur grain
See above, but this time the spirit is added on fermenting grape must and berries rather than only on the must.

Négociant
A producer who buys in grapes or wine and bottles under their own label.

Noble rot (*Botrytis cinerea*)
A good fungus that develops on grape berries, concentrating their sweetness. Noble rot is also responsible for complex aromas in sweet-wine production.

Oxidation
When wine or must is exposed to too much oxygen, it can spoil and develop pronounced nutty and caramelized notes.

Permaculture
*Perma*nent sustainable agri*culture* that seeks to develop self-sufficient ecosystems.

Pressurage direct
When grapes are pressed directly without any prolonged skin contact

Qvevri/Kvevri
Spelled interchangeably, *qvevris* are large clay pots buried underground and used for the fermentation and maturation of wines in traditional Georgian winemaking.

Re-fermentation
When residual fermentable sugars start to ferment again in the bottle.

Reverse osmosis
A very sophisticated, high-tech, selective wine filtration system that can remove volatile acidity, water, alcohol, smoke taint, etc.

Ropiness
Occasionally, during maturation or once bottled, wine can go through a stage where bacteria render the wine oily in texture.

Sterile-filtration
Filtering wine so tightly (down to .45 Q) that yeast and bacteria are eliminated

Stilbene
Naturally occuring antioxidants in wine. Resveratrol is a stilbene.

Sulfites
Wine additives widely used for their antioxidant and antibacterial effects, among others.

Tannin
Naturally contained in grape stems, pips, and skins. Contribute to the sensation of astringency in wines (think strong black tea). Can also be extracted from oak during winemaking.

Tartrate crystals
Also known as cream of tartar. This is the potassium acid salt of tartaric acid. Also known as wine diamonds.

***Teinturier* grape variety**
Literally meaning dyer grape. Red-fleshed grape varieties that produced deeply colored wines.

Tinaja
Spanish clay jar used for the fermentation and maturation of wine.

Vigneron (vigneronne)
Wine producer

Vin liquoreux
Sweet wine

Vintage variation
Growing conditions that are different year to year.

Viticulture (viniculture)
The science of farming vines (specific to grape growing).

FURTHER EXPLORING & READING

GROWER ASSOCIATIONS

Association des Champagnes Biologiques: www.champagnesbiologiques.com

Association des Vins Naturels: lesvinsnaturels.org

PVN (*Productores de Vinos Naturales*): vinosnaturales.wordpress.com

Renaissance des Appellations: renaissance-des-appellations.com

S.A.I.N.S.: vins-sains.org

Taste Life: schmecke-das-leben.at

Vini Veri: viniveri.net

VinNatur: vinnatur.org

Vi.Te (Vignaioli e Territori): vignaiolieterritori.it

WINE FAIRS

RAW WINE (London | Berlin | New York | Los Angeles| Montréal): rawwine.com

Les 10 Vins Cochons: les10vinscochons.blogspot.com

À Caen le Vin: vinsnaturelscaen.com

Les Affranchis: les-affranchis.blogspot.com

Buvons Nature: buvonsnature.over-blog.com

La Dive Bouteille: diveb.blogspot.co.uk

Festivin: festivin.com

Les Greniers Saint Jean: renaissance-des-appellations.com

H_2O Vegetal: h2ovegetal.wordpress.com

Les Pénitentes: www.facebook.com/LesPenitentesAtLeGouverneur/

La Remise: laremise.fr

Real Wine Fair: therealwinefair.com

Salon des Vins Anonymes: vinsanonymes.canalblog.com

Vella Terra: vellaterra.com

Vini Circus: vinicircus.com

Vini di Vignaioli: vinidivignaioli.com

VinNatur Annual Tasting: vinnatur.org

STOCKISTS

Since the first edition of this book, stockists of natural wine have exploded in number around the world, be they fine dining establishments, casual eateries, bars, stores, or even importers and distributors. RAW WINE's People & Places pages are full of suggestions. To explore these, visit: rawwine.com/people-places

RECOMMENDED BOOKS

These are books that I have read and which may be of interest to you as well. They are not necessarily related to natural wine, but many are building blocks that have helped inform what I do and why. Happy reading.

Abouleish, Ibrahim, *Sekem: A Sustainable Community in the Egyptian Desert* (Floris Books, 2005)

Augereau, Sylvie, *Carnet de Vigne Omnivore—3e Cuvée* (Hachette Pratique, 2010)

Allen, Max, *Future Makers: Australian Wines For The 21st Century* (Hardie Grant Books, 2011)

Bird, David, *Understanding Wine Technology: The Science of Wine Explained* (DBQA Publishing, 2005)

Bourguignon, Claude & Lydia, *Le Sol, la Terre et les Champs* (Sang de la Terre, 2009)

Campy, Michel, *La Parole de Pierre—Entretiens avec Pierre Overnoy, vigneron à Pupillin, Jura* (Mêta Jura, 2011)

Corino, Lorenzo, *The Essence of Wine and Natural Viticulture* (Quintadicopertina, 2018)

Chauvet, Jules, *Le vin en question* (Jean-Paul Rocher, 1998)

Columella, *De Re Rustica: Books I–XII* (Loeb Classical Library, 1989)

Diamond, Jared, *Collapse* (Penguin Books, 2011)

Feiring, Alice, *Naked Wine: Letting Grapes Do What Comes Naturally* (Da Capo Press, 2011)

Goode, Jamie and **Harrop**, Sam, *Authentic Wine: toward natural sustainable winemaking* (University of California Press, 2011)

Gluck, Malcolm, *The Great Wine Swindle* (Gibson Square, 2009)

Jancou, Pierre, *Vin vivant: Portraits de vignerons au naturel* (Editions Alternatives, 2011)

Joly, Nicolas, *Biodynamic Wine Demystified* (Wine Appreciation Guild, 2008)

Juniper, Tony, *What Has Nature Ever Done For Us? How Money Really Does Grow On Trees* (Profile Books, 2013)

Mabey, Richard, *Weeds: The Story of Outlaw Plants* (Profile Books, 2012)

Matthews, Patrick, *Real Wine* (Mitchell Beazley, 2000)

McGovern, Patrick E., *Ancient Wine: The Search for the Origins of Viniculture* (Princeton University Press, 2003)

Morel, François, *Le Vin au Naturel* (Sang de la Terre, 2008)

Pliny (the Elder), *Natural History: A Selection* (Penguin Books, 2004)

Pollan, Michael, *Cooked: A Natural History of Transformation* (Penguin, 2013)

Robinson, Jancis, **Harding**, Julia and **Vouillamoz**, José, *Wine Grapes* (Penguin, 2012)

Thun, Maria, *The Biodynamic Year—Increasing yield, quality and flavor, 100 helpful tips for the gardener or smallholder* (Temple Lodge, 2010)

Waldin, Monty, *Biodynamic Wine Guide 2011* (Matthew Waldin, 2010)

WEBSITES AND BLOGS

For Patrick Rey's Mythopia Series, visit capteurs-de-nature.com/Z/Mythopia/index.html

The following writers regularly cover natural wine in some way (apologies if I have left anyone out):

alicefeiring.com (US author and journalist)

blog.lescaves.co.uk (UK importer/distributor/retailer)

caulfieldmountain.blogspot.com (Australian journalist and author)

dinersjournal.blogs.nytimes.com/author/eric-asimov (Eric Asimov, journalist and critic for NY Times)

glougueule.fr (French journalist and activist)

ithaka-journal.net (ecology and wine—with articles by Hans-Peter Schmidt)

jimsloire.blogspot.co.uk (UK investigative blogger)

louisdressner.com (US importer)

montysbiodynamicwineguide.com (UK biodynamic consultant and author)

saignee.wordpress.com (blogger)

vinosambiz.blogspot.co.uk (Spanish natural producer and regular blogger)

wineanorak.com (UK author and blogger)

winemadenaturally.com (UK journalist)

wineterroirs.com (French blogger and photographer)

wine-searcher.com (good source of general wine news)

INDEX

Page numbers in *italics* refer to captions

ACKNOWLEDGMENTS

Author's Acknowledgments

First off, big thanks are due to Cindy Richards and the team at CICO for giving me the chance to write this book. Thank you for your relentless patience, diligence, and for not giving up (especially Penny Craig, Caroline West, Sally Powell, and Geoff Borin who have done more than their fair share of waiting). Thanks also to Matt Fry for opening the door, and to Gavin Kingcome for his wonderful photography.

Keen thank yous are also due to Dr. Laurence Bugeon and Fränze Progatzky for getting me excited about microscopes, to Marie Andreani for giving up her time to help with transcriptions of interviews, and to all my friends and family who did not see me for months.

But, most of all, thank you to all those of you who gave up your time to share your thoughts and wisdom with me and tell me your stories. A few of you even helped check parts of the final manuscript or shared your photos, some of which are included in this book. A huge thank you particularly to Hans-Peter Schmidt whose time and knowledge have been invaluable.

Finally, last and certainly not least, thank you to my partner Deborah Lambert, without whom this book would never have seen the light of day. Thank you for helping bring order to my unruly thoughts and for making it all sound a little less French!

Publisher's Acknowledgments

The publishers would like to thank the following people for allowing their vineyards, bars, and restaurants to be photographed for the purposes of this book:

France

Alain Castex and Ghislaine Magnier, formerly of Le Casot des Mailloles, Roussillon

Anne-Marie and Pierre Lavaysse, Le Petit Domaine de Gimios, Languedoc

Antony Tortul, La Sorga, Languedoc

Didier Barral, Domaine Léon Barral, Languedoc-Roussillon

Gilles and Catherine Vergé, Burgundy

Jean Delobre, La Ferme des Sept Lunes, Rhône

Jean-Luc Chossart and Isabelle Jolly, Domaine Jolly Ferriol, Roussillon

Julien Sunier, Rhône

Mathieu Lapierre, Domaine Marcel Lapierre, Beaujolais

Pas Comme Les Autres, Bézier, Languedoc-Roussillon

Romain Marguerite, Via del Vi, Perpignan

Tom and Nathalie Lubbe, Domaine Matassa, Roussillon

Yann Durieux, Recrue des Sens, Burgundy

Italy and Slovenia

Aleks and Simona Klinec, Kmetija Klinec, The Brda, Slovenia

Angiolino Maule, La Biancara, Veneto, Italy

Daniele Piccinin, Azienda Agricola Piccinin Daniele, Veneto, Italy

Stanko, Suzana, and Saša Radikon, Radikon, Friuli Venezia, Giulia, Italy

California

Chris Brockway, Broc Cellars, Berkeley

Darek Trowbridge, Old World Winery, Russian River Valley

Kevin and Jennifer Kelley, Salinia Wine Company, Russian River Valley

Lisa Costa and D.C. Looney, The Punchdown, San Francisco,

Phillip Hart and Mary Morwood-Hart, AmByth Estate, Paso Robles

Tony Coturri, Coturri Winery, Glen Ellen

Tracy and Jared Brandt, Donkey & Goat, Berkeley

Picture Acknowledgments

The publishers would like to thank the following people for kindly allowing their photographs to be reproduced:

KEY: t = top b = bottom c = center r = right l = left

Alamy: 124b; **Antidote**: 124tr; **Casa Raia**: 93b; **Château La Baronne**: 29t; **Frank Cornelissen**: 37; **Costadilà**: 138; **Coulée de Serrant**: 42, 44–45 (both); **Domaine de Fontedicto**: 106–107 (both); **Domaine Henri Milan**: 69; **Elliot's**: 124tl; **Gelveri**: 168; ***Hibiscus***: 125; **Nicolas Joly**: 42, 44–45 (both); **Katy Koken**: 158; **Isabelle Legeron MW**: 26tr, 29bl, 33, 35, 37, 43, 47 (both), 79, 82, 88–89 (both), 92b, 96–97, 101, 104 (both), 105, 108–109, 110, 111, 114, 117, 118, 122, 123, 140, 147r, 169, 172, 180r, 184, 194, 200, 203; **Lous Grezes**: 194; **Mamaruta**: 173; **Margins Wine**: 157; **Matassa** (Tom Lubbe and Craig Hawkins): 25; **Montesecundo**: 186; **Patrick Rey** (at Mythopia): 16, 30–31 (all), 32, 98, 189; **Strohmeier**: 34; **Taubenkobel**: 125; **Viniologi**: 167: **Weingut Werlitsch**: 26b, 155